Auto
Electronics
Projects

The Maplin series

This book is part of an exciting series developed by Butterworth-Heinemann and Maplin Electronics Plc. Books in the series are practical guides which offer electronic constructors and students clear introductions to key topics. Each book is written and compiled by a leading electronics author.

Other books published in the Maplin series include:

Computer Interfacing	Graham Dixey	0 7506 2123 0
Logic Design	Mike Wharton	0 7506 2122 2
Music Projects	R A Penfold	0 7506 2119 2
Starting Electronics	Keith Brindley	0 7506 2053 6
Audio IC Projects	Maplin	0 7506 2121 4
Video and TV Projects	Maplin	0 7506 2297 0
Test Gear & Measurement	Danny Stewart	0 7506 2601 1
Integrated Circuit Projects	Maplin	0 7506 2578 3
Home Security Projects	Maplin	0 7506 2603 8
The Maplin Approach to Professional Audio	T.A. Wilkinson	0 7506 2120 6

Auto
Electronics
Projects

Newnes
An imprint of Butterworth-Heinemann Ltd
Linacre House, Jordan Hill, Oxford OX2 8DP

A member of the Reed Elsevier group

OXFORD   LONDON   BOSTON
MUNICH   NEW DELHI   SINGAPORE   SYDNEY
TOKYO   TORONTO   WELLINGTON

British Library Cataloguing in Publication Data
A catalogue record for this book is available from the
British Library
ISBN 0 7506 2296 2

Library of Congress Cataloguing in Publication Data
A catalogue record for this book is available from the
Library of Congress

 Edited by Co-publications, Loughborough

 Typeset and produced by Sylvester North, Sunderland

all part of The Sylvester Press

Printed in Great Britain by Clays Ltd, St Ives plc

Contents

Preface

This book is a collection of articles and projects previously published in *Electronics — The Maplin Magazine*.

Each project is selected for publication because of its special features, because it is unusual, because it is electronically clever, or simply because we think readers will be interested in it. Some of the devices used are fairly specific in function — in other words, the circuit is designed and built for one purpose alone. Others, on the other hand, are not specific at all, and can be used in a number of applications.

This is just one of the Maplin series of books published by Newnes books covering all aspects of computing and electronics. Others in the series are available from all good bookshops.

Maplin Electronics Plc supplies a wide range of electronics components and other products to private individuals and trade customers. Telephone: (01702) 552911 or write to Maplin Electronics, PO Box 3, Rayleigh, Essex SS6 8LR, for further details of product catalogue and locations of regional stores.

1 Car electrical systems

The modern motor vehicle is a precision-built highly-tuned machine. High speed performance, low fuel consumption and quiet smooth-running engine all rely on efficient ignition, battery charging and general electrical systems throughout the car.

The electrical system is very complex. One only has to look behind a dashboard to see the hundreds of wires of all sizes and colours, interconnecting the instruments, high voltage and high current circuits. Also, the electrical system is very prone to breakdown, whether this is a blown lamp bulb, a faulty dynamo or badly adjusted contact breaker points.

Auto electronics projects

No two models of cars have identical electrical circuits. The electrical circuits are, however, similar and fall into categories such as conventional ignition or electrical ignition, dynamo or alternator, positive or negative earth.

This chapter describes the basic systems: it is left to the individual car owner to interpret the descriptions and diagrams to suit their particular vehicle.

One word of warning. Car electric circuits can cause damage to either the car or to the user if tampered with. For instance a short circuit across the battery can generate hundreds of amperes and a lot of heat, even a fire: the ignition circuit generates very high voltages indeed: tampering with the instrument circuits, can cause misleading readings and a possible safety hazard to the driver. Before embarking on any changes to the car electrics, make every effort to understand how the circuit works. In this way fault finding should be greatly simplified.

The ignition circuit

The purpose of the ignition circuit (Figure 1.1) is to supply the high voltage required to operate the spark plugs in the correct sequence and so ignite the air/petrol mixture in each cylinder. The explosions generated push the pistons and so turn the engine, causing motion. The circuit comprises the car battery, an ignition coil, the distributor and four (or six) spark plugs. The principle of operation is described later.

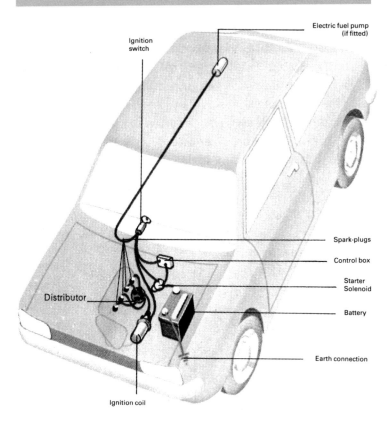

Figure 1.1   The ignition circuit

Battery charging

All electrical systems draw their power from the 12 volt
battery (Figure 1.2). If the battery was not continually
charged it would become exhausted very quickly, particularly if the lights, wipers and starter motor were in
constant use. The turning of the engine charges the battery by connecting it to a dynamo, via the fan belt. A

Auto electronics projects

pulley network at the front of the engine constantly turns the dynamo which generates enough power to charge up the battery. A control box controls the charging rate and informs the driver via the ignition light if the battery is not charging. Some cars use an alternator in preference to a dynamo. These are more efficient but generate a.c. rather than d.c. and so require rectification of the a.c. output. Battery charging is described later.

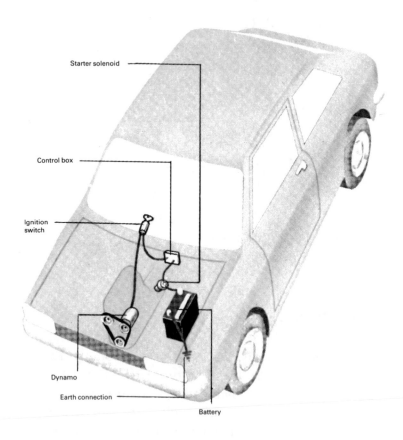

Starter solenoid

Control box

Ignition switch

Dynamo

Earth connection

Battery

Figure 1.2   The battery charging circuit

Lighting

The lighting circuits are the simplest of all these, comprising a simple connection of the 12 volt lamp to the battery via the instrument panel switches (Figure 1.3). These circuits are completely independent of the ignition and charging circuits, the one connection to each lamp being taken via a single wire and respective switch to the battery; the other connection uses the car chassis. The lighting circuits are described in more detail later.

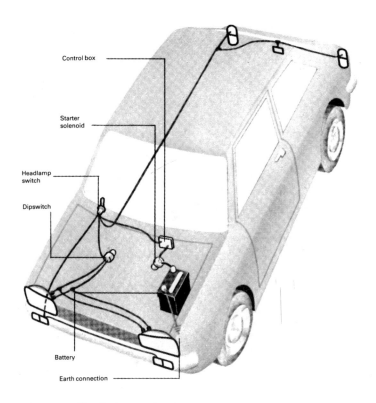

Control box

Starter solenoid

Headlamp switch

Dipswitch

Battery

Earth connection

Figure 1.3 The lighting circuit

Auto electronics projects

Indicators and accessories

Contained within this circuit is the starter motor which draws hundreds of amperes from the battery to turn the engine until it fires (Figure 1.4). Heavy duty cable and a heavy duty solenoid carry out this operation, which is prone to trouble for various reasons. Also there is the fuel pump which is a small solenoid operated device to

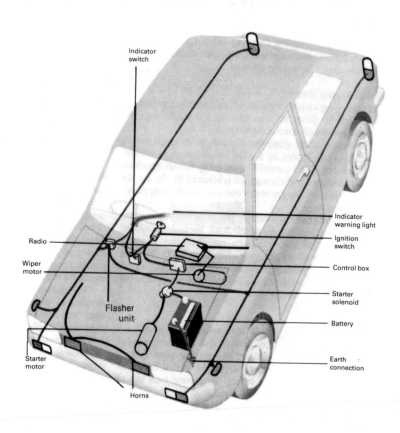

Figure 1.4   The indicator and accessories circuit

pump petrol from the tank to the carburettor, the indicator light circuitry with hazard warning lights, the radio and cassette player circuits, the heater and wiper motors, horns, instrument gauges, and heated rear screen. These circuits are relatively simple and are described together with fault-finding techniques later.

Wiring diagram

Car wiring diagrams are often very difficult to read and interpret. The reason for this is that, in a modern car with a large number of instruments, lights, accessories and motors, all are to be interconnected on one comprehensive diagram. Fuses and switches must also be shown, together with the colours of the wires and cables; many manufacturers use an international colour code for easier identification of the respective circuit cables.

Some of the more popular symbols used in car wiring diagrams are illustrated in Figure 1.5. The cables are often coded and coloured for identification and a shorthand method of simplifying the diagram often groups all in one bundle (called a cable-form) as a single line. To trace the start and finish of one cable involves almost microscopic analysis of all connections, searching for the required code and colour.

Electronic devices such as electronic ignition or the dashboard microprocessor are shown as simple blocks. Fault finding within these devices must be left to the specialist dealer.

Auto electronics projects

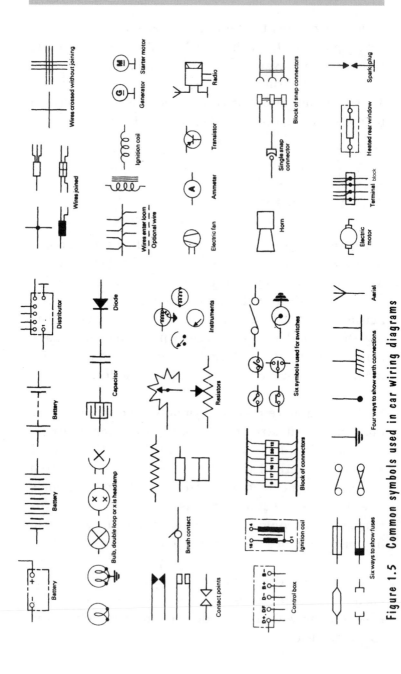

Figure 1.5 Common symbols used in car wiring diagrams

8

The engine

The most common small to medium car engine is the 4-cylinder petrol internal combustion engine. More powerful engines have six cylinders, some have eight; motor cycles and mopeds have one or two. The arrangement of cylinders varies, some being overhead cam shaft, some pushrod and rocker, and others with cylinders aligned in the shape of a *V*.

This brief description of the 4-cylinder engine, highlights the importance of accurate timing so as to maximise power and performance. Figure 1.6 shows the arrangement of cylinders and the four strokes, illustrated separately in Figure 1.7:

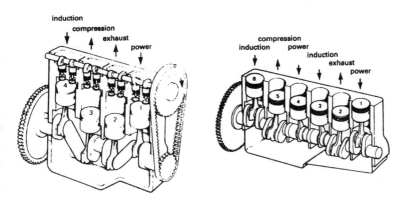

Crank position (degrees)	Cylinder no. 1	Cylinder no. 2	Cylinder no. 3	Cylinder no. 4
0–180	Power	Exhaust	Compression	Induction
180–360	Exhaust	Induction	Power	Compression
360–540	Induction	Compression	Exhaust	Power
540–720	Compression	Power	Induction	Exhaust

Figure 1.6   4-cylinder and 6-cylinder engines

Auto electronics projects

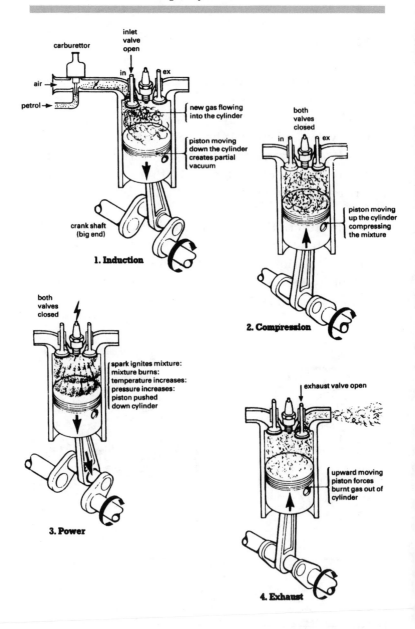

Figure 1.7   The four stages of combustion

●   induction — the petrol/air mixture is sucked into the cylinder,

●   compression — the piston compresses the mixture,

●   power — the spark plug ignites the mixture causing an explosion which pushes the piston down,

●   exhaust — the piston pushes the burnt gases out of the cylinder.

The four cylinders operate in series so that, at any one time, one is being powered. The crank shaft positions the pistons in the correct sequence, two complete revolutions (720°) comprising the complete four-stroke cycle. The electrical circuits have the job of supplying each spark plug with a high voltage pulse to power the piston in the correct sequence, and at the time when the piston is at the top of its stroke (top dead centre). The distributor ensures that the pulses travel in sequence to the four spark plugs and, at the same time, time the pulse to top dead centre.

Basic ignition

The main components of the ignition circuit are the ignition coil — a cylindrical transformer with two connections SW and CB and a high tension cable going to the distributor (see Figure 1.8) — and the distributor — a mechanical device coupled to the engine via skew gears. This acts as a four-way switch to route the high tension to the spark plugs, and as a means of generating the high tension voltage.

Auto electronics projects

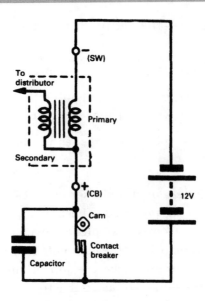

Figure 1.8   Basic high voltage generating circuit

Figure 1.8 shows the basic high voltage generating circuit. The operation is as follows, assuming the contact
breaker points are initially closed (see Figure 1.10):

●   the piston in one cylinder (say number 1) rises to
top dead centre, compressing the petrol/air mixture,

●   the rotor arm in the distributor cap points to the
appropriate high tension connection to spark plug
number 1 and,

●   the contact breaker points open,

●   the magnetic field in the primary of the ignition coil
(Figure 1.9) quickly collapses. The turns ratio of the
transformer of about 10,000 to 1 transforms this collapse
into a voltage of about 20,000 volts across the secondary,

12

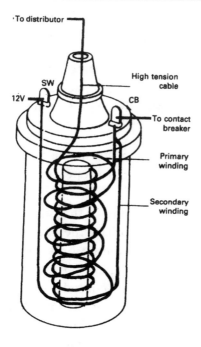

Figure 1.9   The ignition coil

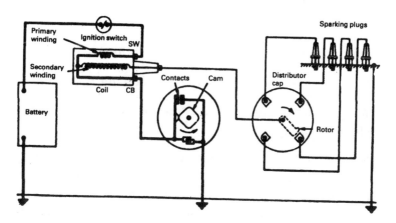

Figure 1.10   Sparking plugs firing circuit

13

Auto electronics projects

●   the high tension pulse ignites the petrol/air mixture in cylinder 1 causing the engine to rotate,

●   the distributor shaft rotates to again close the contact breaker points. The capacitor across the points suppresses the high voltage pulse generated by this closure,

●   the distributor shaft turns the rotor arm to the next cylinder and the procedure repeats.

The timing of the opening of the points is critical. The distributor shaft cam opens the gap as in Figure 1.12, the positioning of the contact breaker points assembly is critical together with the gap width. The points, after a period of wear, tend to corrode and pitting occurs; a deposit which builds up and reduces the effective gap. The gap is usually about 25 thousands of an inch wide, opens and closes some ten million times every 1000 miles. One other adjustment to optimise the timing is the dwell angle. This is the number of degrees that the points remain closed; refer to the maker's manual for the recommended value.

Ignition timing is carried out in the following sequence:

●   choose cylinder number 1 — consult the manual,

●   locate the timing marks on the fan belt pulley (see Figure 1.13),

●   turn the engine crank shaft until the marks align at top dead centre (t.d.c.). The engine can be turned by placing the car on level ground, take out all the spark plugs, place in top gear, release the brakes and move the car to and fro,

● ensure that the distributor rotor arm points to the high tension lead to cylinder number 1. If not, turn the engine through a further 360°,

● connect a 12 V lamp between the contact breaker spring (see point X in Figure 1.12) and a good earth point,

● rotate the engine by about 20°, then inch it slowly backwards until the lamp just lights,

● if the t.d.c. reading is incorrect, align the t.d.c. mark, then loosen the distributor clamping nut (point Y in Figure 1.11) and turn the entire distributor anticlockwise until the light just goes out. Then turn clockwise until it just lights. Clamp the nut,

● check the t.d.c. setting once again,

● replace the plugs, put on the brakes and take out of gear! A faster method uses a stroboscope with the engine running, a Xenon tube flashing as the points open and close.

Electronic timing

The system so far described sometimes fails because of pitting of the points and wear and tear of the moving parts of the distributor. Two types of electronic system are found:

● transistorised ignition or capacitor discharge ignition — see Figure 1.14 and,

● contactless (optical or magnetic) ignition.

Auto electronics projects

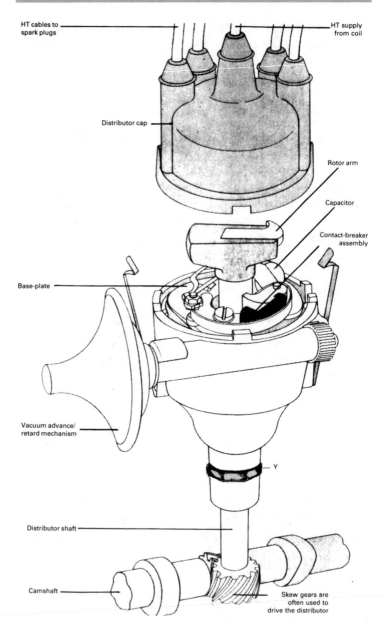

HT cables to spark plugs

HT supply from coil

Distributor cap

Rotor arm

Capacitor

Contact-breaker assembly

Base-plate

Vacuum advance/ retard mechanism

Y

Distributor shaft

Camshaft

Skew gears are often used to drive the distributor

Figure 1.11   The distributor

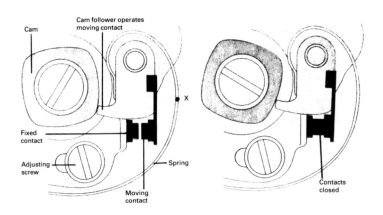

Figure 1.12   Contact breaker assembly

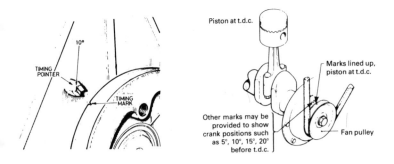

Figure 1.13   Timing marks on fan belt pulley

Transistor ignition uses a power d.c.–d.c. converter, a two transistor push-pull oscillator, to generate 400 V or so, to feed to the ignition coil and produce a higher voltage and healthier spark. At the same time, the contact breakers no longer switch the full 12 volt battery current: they merely switch a 12 volt low current signal to the d.c.–d.c. connector. The points therefore last far longer and the system is virtually maintenance-free.

Auto electronics projects

Contactless ignition uses a moving magnet or infra-red ray to replace the cumbersome contact breakers, a transistorised d.c.–d.c. converter circuit being used as before to deliver the high tension pulses to the plugs. Both systems can be installed into an existing circuit in a very small time, a number of modern cars having such systems built in when new.

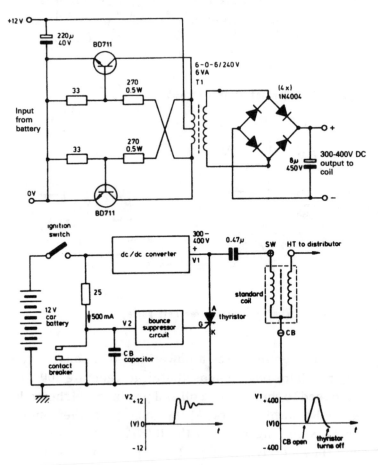

Figure 1.14   Transistorised and capacitor-discharge ignition circuits

The battery

A car battery is a real powerhouse and should always be maintained in prime condition. It is comprised of a series of six lead-acid 2 volt cells (Figure 1.15) which, together, constitute 12 volts at capacities varying from about 30 to 100 ampere-hours. A 70 ampere-hour battery delivers a constant 70 amps for one hour, or one amp for 70 hours, or on a very cold day, 400 amps for a few seconds to start the engine.

The negative plates are constructed from spongy lead plates and the positive plates from lead dioxide. Dilute sulphuric acid with a specific gravity of about 1.2 starts the chemistry into action, current from the battery turning the plates into lead sulphate. A battery charger, by

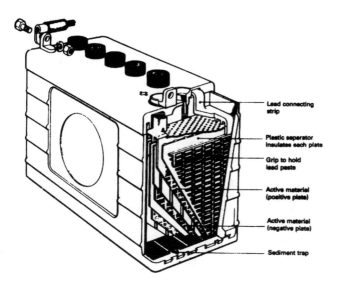

Lead connecting strip

Plastic separator insulates each plate

Grip to hold lead paste

Active material (positive plate)

Active material (negative plate)

Sediment trap

Figure 1.15   The battery

way of the dynamo or alternator, reverses this process by restoring the battery plates to their original composition.

Modern batteries are self maintaining and the electrolyte (acid) levels remain constant. Older batteries are prone to deterioration and last only 3 or 4 years. The performance of a battery falls at low temperatures, giving problems on a cold morning and sulphation of the terminals which causes leakage currents to chassis; this is avoided by smearing petroleum jelly onto the terminals. A more common cause of battery trouble, other than an old and tired battery itself, is damp and dirty wiring, particularly around the starter motor which drains most of the battery power.

Battery charging is carried out in one of two ways:

● the dynamo — a d.c. generator, like a motor in reverse, which delivers current to the battery as long as the engine is running fast,

● the alternator — an a.c. generator which, although requiring an a.c./d.c. rectifier circuit, has greater efficiency and charges the battery even when idling.

Figure 1.16 shows a cut away picture of the dynamo and the circuit which controls the charging of the battery, called the cut-out or control box. This unit senses the dynamo output voltage and, if low, cuts the dynamo out of circulation. As the voltage rises the cut-out connects the dynamo to charge the battery and if it rises beyond a preset value, the regulator winding reduces the effective dynamo output by adjusting the current in the field winding, excessive current going directly to the car electrical circuits.

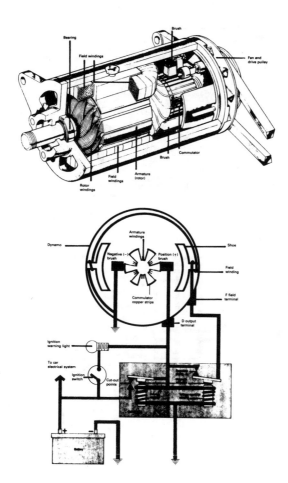

Figure 1.16   Dynamo and control box

The alternator is shown in Figure 1.17 together with its control circuitry and rectifier diodes. The three stator windings are connected internally to the diodes and a d.c. output is obtained. A transistorised control circuit maintains a constant battery charging current by adjusting the current in the rotor winding.

Auto electronics projects

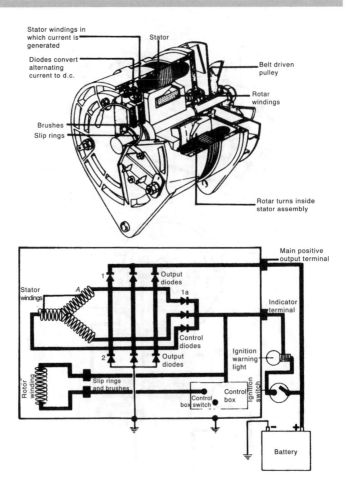

Figure 1.17 Alternator and control circuitry

Both systems have a built-in ignition warning light with one side connected to the battery +12 V terminal, the other to the dynamo or alternator output. If the generator is not working, when the engine is switched off for

Car electrical systems

instance, or when the fan-belt is slipping or broken, the 12 V bulb has 12 volts acrʌss it and it lights. Normally the lamp has 12 volts on either side and it goes out.

Lighting

Little needs to be said about the normal lighting circuits except to say that the headlamp bulbs can consume several amperes each and so cable of the correct size must be used to prevent heating (or melting) of the wiring. Many bulbs, as in Figure 1.18, have two filaments for compactness. Quartz halogen bulbs, with a gas surrounding the tungsten filaments, give off greater brightness.

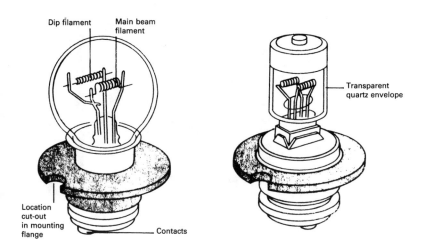

Figure 1.18   Dual filament bulbs

Auto electronics projects

As the headlamps between them consume several amperes, the headlamp (or flasher) switch has to be heavy duty and high current wires must be sent to the dashboard. Consequently a relay is often positioned near the headlamps, as in Figure 1.19, this being activated via a (preferred) low current switch and wiring. Operating the switch activates the relay which connects the headlamps directly to the battery terminal.

One final lighting device in common use is the spring steel flasher unit (see Figure 1.20) which turns the indicator lamps on and off.

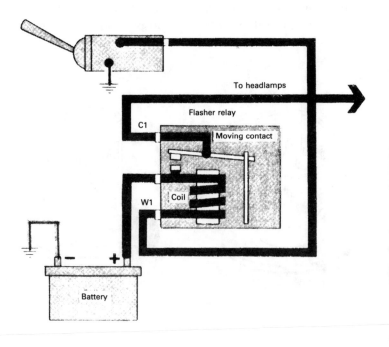

To headlamps

Flasher relay

C1

Moving contact

Coil

W1

Battery

Figure 1.19   Headlamp relay

While cold, the contacts are held together by the diaphragm. When current passes through the contacts, by indicating to turn left or right, the resistance metal heats up, expands and pushes the contacts apart. They then cool again, close and the sequence repeats 60 to 120 times a minute. Emergency light units are similar except that heavy duty contacts are used.

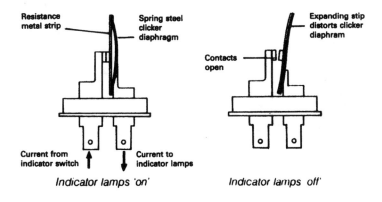

Figure 1.20   Flasher unit

Starter motor and other accessories

In a similar way to the headlights being operated via a *remote control* relay, a starter solenoid is used as in Figure 1.21 to switch the 400 amps to the starter motor. This wiring is the thickest to be seen under the bonnet and every step is taken to minimise any heat generated despite the costs of the thick copper wire. The starter motor engages with the engine via the flywheel to start the engine, as seen in Figure 1.22. If the ignition circuit is working well, a few turns of the engine should cause the engine to fire and continue under its own steam. The starter motor is then disconnected from the engine.

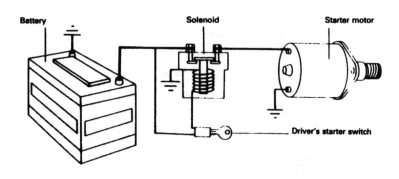

Figure 1.21   Starter solenoid

Two methods are used, a pre-engaged motor whose pinion is always linked to the flywheel, a solenoid operating a plunger to engage the starter motor with its pinion (like a small clutch), and the inertia type whose pinion slides along the shaft to engage with the flywheel as soon as the starter motor operates. These are shown in Figure 1.23. Figures 1.24 to 1.28 illustrate a number of other electrical accessories which are essential, and some legally required, in the modern motor car.

26

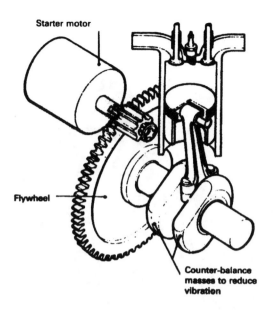

Figure 1.22   Flywheel

Petrol pumps operate either via a mechanical rocker assembly coupled to the engine forming a small mechanical pump (Figure 1.24), or an electrical diaphragm pump, rather like a vibrator, which pumps the petrol from the tank to the engine, as in Figure 1.25. The petrol gauge operates using a small float coupled to a variable resistance unit. As the petrol level rises or falls, the current to the gauge rises or falls accordingly. This unit, similar to a WC ball-cock, is sealed for fire reasons, see Figure 1.26.

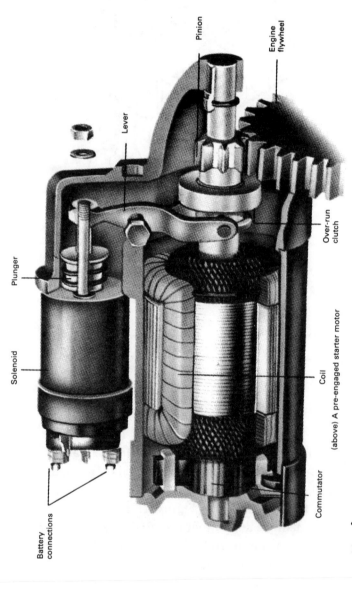

(above) A pre-engaged starter motor

Figure 1.23   Starter motors

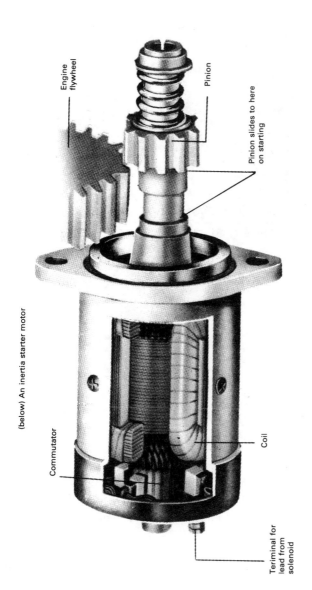

(below) An inertia starter motor

Engine flywheel

Pinion

Pinion slides to here on starting

Commutator

Coil

Teriminal for lead from solenoid

Figure 1.23 Continued

Auto electronics projects

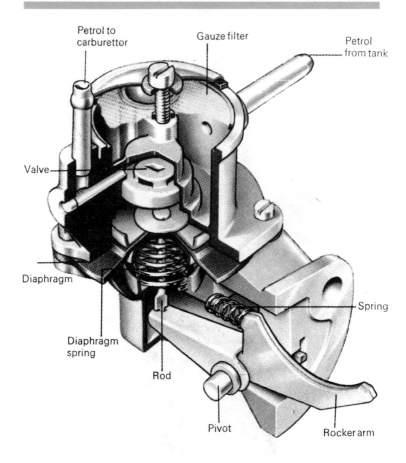

Figure 1.24   Mechanical fuel pump

Horns come in all shapes and sizes. Figure 1.27 shows a simple type, working like a vibrator whose diaphragm output is mechanically amplified to warn pedestrians to get out of the way.

Ammeters can be fitted in any car: a simple means of installation necessitating a minor change to the wiring

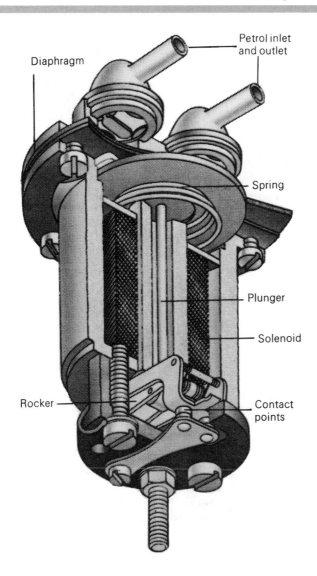

Diaphragm

Petrol inlet
and outlet

Spring

Plunger

Solenoid

Rocker

Contact
points

Figure 1.25   Electric fuel pump

Auto electronics projects

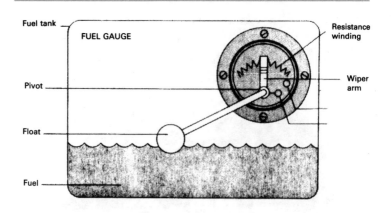

Figure 1.26 Fuel gauge and float

as shown in Figure 1.28. By this means the ammeter does not record the starter motor current, but all other currents taken by the car circuitry.

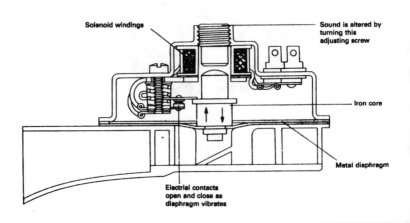

Figure 1.27 Horn diaphragm

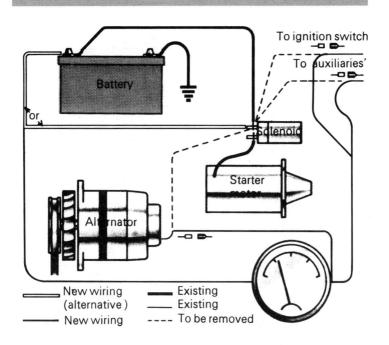

To ignition switch

To auxiliaries'

Battery

or

Solenoid

Starter motor

Alternator

	New wiring (alternative)		Existing
	New wiring		Existing
			To be removed

Figure 1.28   Ammeter wiring

Finally, a look into the computerised dashboard now found in a number of high performance cars. Transducers constantly read r.p.m., pressures, temperatures and so on; these are monitored and the computer checks and warns the driver of impending trouble (see Figure 1.29). The day of the James Bond supercar or the Night Rider's *Kit* looms nearer everyday.

Auto electronics projects

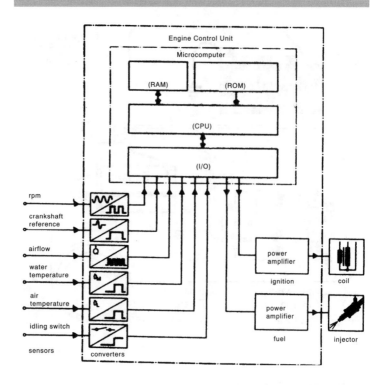

Figure 1.29   Computerised dashboard

2 Electronic ignition

The electro-mechanical ignition system that has been used to fire the fuel/air mixture in an internal combustion engine for several decades, and which is familiar to home mechanics everywhere, has practically been replaced by electronic methods in recent times. Some of the reasons for this are not quite as obvious as you might suppose, but certainly, as with everything else, a modern electronic alternative is superior to its electro-mechanical ancestor. To be fair though, the latter has had a lot going for it, it originally replaced a method so archaic as to be unbelievable.

Automotive ignition — a brief history

Earliest motor cars, or in fact anything using the new-fangled *gas* engine (many of which were also used for powering agricultural machinery), of slightly over a century ago had to make do with a device comprising a thin-walled copper tube with closed ends, supported in the middle with a porcelain insulator or some-such similar item. The insulator screwed into the cylinder head, like a modern plug — indeed the word *plug* probably originates from this time.

To start the engine, the outside end of the tube is heated with the flame of a spirit burner until glowing. Then attempts can be made to get the engine going, using a starting-handle. When the fuel/air mixture arrives at the other end of the tube, on the inside, in the right quantities (a bit of a juggling act), it should (hopefully!) burn. Once the engine is warmed up and running, the spirit burner can be put out and thereafter the temperature of the tube will be maintained by the heat of internal combustion, in the same way that the engine of a model aeroplane keeps its *glow-plug* hot.

Not surprisingly, while the *gas* engine was still only a few years young, engineers thought hard about improving this less than ideal situation. It was only a question of time before the electrically powered *hot wire* type of ignition, a *glow-plug* then, was pressed into service for the petrol engine. The trouble with glow-plugs however, is that the wire burns away quite quickly and a stock of spares must be carried around at all times.

Then, just prior to the turn of the century, a method was devised which, though it seems obvious now, must have taken a good deal of working out at the time. It was reliable in operation like nothing else previously, it was sophisticated, it was state-of-the-art. It was spark ignition.

The advantages included much easier starting — simply energise the system and crank the handle. Also, because the plug was no more than a spark gap at the *business end*, and the electrodes were far more robust than thin wire or copper tube, it had a working life hitherto unseen.

From the engine designers' point of view it raised two important possibilities:

● the moment of ignition of the fuel/air mixture could be precisely controlled. Previously, the combustion chamber had to be designed to prevent the *charge* igniting prematurely during compression, a shape which did nothing for efficiency (or *performance*, if you like),

● engines with multiple cylinders could be catered for just as easily as singles. Prior to this engines were mostly a single cylinder type — the ignition paraphernalia for just one was usually quite enough to cope with.

There are basically two types of electro-mechanical spark ignition systems: the magneto, and what's called coil ignition. The only difference is that the magneto also generates its own electric power to operate. With coil ignition the power supply is external. In the beginning, there was only the magneto. In the 1920s, the Americans pioneered coil ignition, which used power hitherto generated exclusively for *ancillaries* — lights and so forth. The power supply comprised a d.c. generator in the form of a dynamo, with a *back-up* for the periods when the dynamo couldn't provide the necessary current — an *accumulator* (a battery). In Europe there was great resistance to coil ignition, especially among the British, who thought it *too gimmicky*. Customers wouldn't buy a car if it had coil ignition — manufacturers had to revert to the magneto in order to be able to maintain sales. Would you believe that such a respected manufacturer as Rolls Royce couldn't shift their latest sports tourer until they had put a magneto back into every car? Such was the resistance to change. Perhaps there is a modern parallel here, about customers (and mechanics) being frightened of the complexity of fuel injection...

Spark ignition — the principles

An electric arc is an electric current flowing through a gas, which for the purposes of this discussion, is air. Air, as with most *insulators* resists the flow of electric current. If forced, it *ionises* as electrons begin to move between molecules. As with any other resistor, this molecular friction generates heat — from the amount of energy required to cause air to succumb, quite a lot of heat. The arc is a white/blue colour, and hot enough to start a fire.

It is worth describing how the electro-mechanical ignition system operates first, since there is no substantial difference between it and any electronic equivalent — they all have to do the same thing, make a spark. We shall start here and work backwards.

Air needs a little persuading in order to carry an electric current and produce an arc. At normal atmospheric pressure it is not all that difficult, but still requires a high voltage to *break down* the air between a pair of electrodes. The narrower the gap, the easier it is. However, whilst it is quite easy to bridge a gap of 0.02 inches (a typical spark plug gap) in *open air*, it is much more difficult inside the combustion chamber. This is because air ionises more easily the thinner it is (the typical demonstration is an electric arc in a glass vessel with a vacuum pump attached), it correspondingly becomes more resistive the more dense it is, like inside the combustion chamber of an engine. Universally, the fuel/air mixture is compressed before ignition, the main reason being that this releases more energy on combustion (but also because the piston, being a reciprocating part linked to a revolving part, can't help itself). The upshot of all this is that it is more difficult to bridge the gap to produce a spark in consequence, requiring a very high voltage to do so, which accounts for the 25 to 35 kV HT voltage range typical at the plug's *live* end. I labour on this point because it causes problems for the design of electronic ignition amplifiers, as will be seen later.

Obviously it is impractical for this sort of potential to be produced and controlled directly from some engine driven generator, so instead a step-up transformer is used, which is where the coil comes in. All the generation and timed-switching is done at a more manageable low voltage, and is converted by the coil to the necessary high voltage.

Actually the system is cleverer than that. The sequence shown in Figure 2.1(a) to 2.1(d) reveals the system to be a form of *flyback converter*. Figure 2.1(a) shows the components of a mechanical system *at rest*. With switch S1 open, nothing is happening. When S1 closes in Figure 2.1(b), current flows in the primary winding L1 of T1,

Auto electronics projects

the ignition coil. T1 has a laminated steel core and a finite time is taken for this core to reach magnetic saturation, by which time the primary current will also be at a maximum. This maximum is set by choosing a d.c. impedance for L1 by using resistive wire, or else it will attempt to short-circuit the supply after the core saturates! For 12 V systems the impedance is chosen for a maximum current of around 3.5 to 4 A, as a typical value.

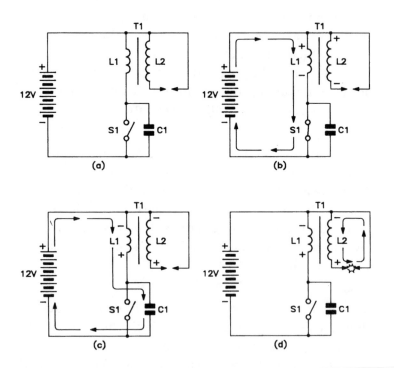

Figure 2.1   Sequence of activity in contact breaker ignition system

In Figure 2.1(c), S1 opens and unwanted effects take place in its vicinity, but we'll ignore them for the moment. Suffice to say that as the magnetic field collapses, it attempts to maintain the current flow in L1 in the same direction, and at the same time induces a current in L2. Because L2 has many more turns than L1, its output voltage is much higher. In the characteristic manner of flyback converters, the coil will attempt to output the same amount of power that went into it. If a path on the primary side is denied it, then the only recourse is to find an outlet on the secondary side.

The *load* is the plug air gap, which basically doesn't want to know at first, but the coil will keep pushing the voltage up until the gap is bridged. If the total power input was 50 W and the output reached 30 kV then the gap current is initially 1.6 mA. However, once the arc is started, the voltage level required to maintain it can reduce substantially allowing a greater current flow and a nice healthy spark. This is indicated in Figure 2.1(d).

The snag is that a smaller representation of this activity also appears across the primary, L1. The effect is an initial pulse of up to several hundred volts. At the point of breaking the circuit, the mechanical switch S1 has a very narrow gap between its contacts which might be measured in microns. Such a gap is easy for a couple of hundred volts to bridge; the coil expends all its energy in producing an arc between the switch contacts, and there is none left for the plug. If you want to prove the effect for yourself try it with the coil of a relay, a pair of test leads and a battery.

So this is where the other clever bit comes in, the third component in the set-up, C1. To this day it is still called a *condenser*, a very old-fashioned name for a capacitor. Its function is to momentarily take over from the switch. As S1 opens, current flow is diverted into C1, charging it. The idea is that by the time the primary voltage reaches a high level, the contact gap is unattainably wide, forcing the coil to go for the plug gap instead. This has two main disadvantages:

● it consumes some power which might otherwise contribute to the spark, and,

● it slows down the rate at which the HT level can increase, the output of which takes on more of a milder ramped pulse shape rather than a true pulse. The value of C1 is critical: if too small, it will encourage switch arcing; if too large, it will absorb too much power and defeat the whole object. A value of 220 nF is usually about right. Switch arcing and power loss still occur, but at acceptable levels.

A third anomaly is that, after the main pulse has occurred, what you are left with is L1 and C1, with the supply as a common terminal, forming a tuned circuit which *rings* or resonates slightly. Figure 2.2 shows the voltage waveforms associated with this series of events.

It was mentioned that the ignition coil has an inbuilt d.c. impedance to limit current flow while the contact breakers are closed. During this time the coil is drawing its maximum power of 45 to 50 watts, to no effect other than that this manifests itself as heat. Consequently an ignition coil has been safeguarded against this, and hence is almost universally constructed as shown in Figure 2.3. It is supported in the centre of an aluminium can, which is filled with oil. An ignition coil is, therefore, a liquid cooled component.

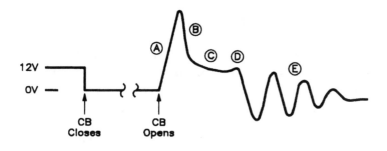

Figure 2.2   Voltage waveform from Figure 2.1 at coil primary

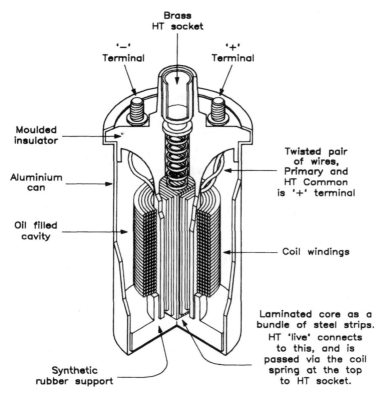

Figure 2.3   Internal construction of a typical ignition coil

Advantages of electronic ignition

The first two problems are practically solved by electronic switching, the third by using the coil in a different way. There are other problems that can be solved at a stroke, like mechanical wear.

The heel of the moving half of a contact breaker wears on the distributor cam. The contact surfaces become damaged, developing a hole or pit in the positive side and a raised *pip* on the negative surface, as the inevitable arcing causes metal to migrate from one surface to the other. The lumpy result causes irregular timing and bad separation, but it may be possible to *rescue* them with the skilful application of a fine stone.

Then there is the (sometimes better than dreadful) mechanical auto-advance mechanism, with its centrifugal bobweights, springs, cam contours and vacuum assist device. To be fair, in practice a mechanical system which is both well designed and 100% fit is difficult to beat, even by an electronic equivalent, but sooner or later wear takes its toll, affecting engine efficiency, and so it needs periodic examination and correction or even replacement.

But owners put off having the car serviced until it desperately needs it because of *exorbitant* garage bills. In the meantime the vehicle is wasting valuable fossil fuel and polluting the atmosphere in a way that it wouldn't if properly tuned. Also of concern to car manufacturers, under pressure to reduce pollution and fuel consumption, is the D.I.Y. home mechanic tinkering with his engine. If he knows what he is doing then fine. If he doesn't...

Consequently factory set and maintenance free electronic ignition, and carburettors with security blanking plugs sealing off the vital bits, prevent unauthorised hands fiddling with these and getting it wrong. And you thought it was all done for your benefit. It also explains the lack of really meaningful information in the modern owner's handbook. *Refer servicing to your dealer, or warranty is void*, and that sort of thing. Basically it means slapped wrist to the potential D.I.Y'er.

Electronic ignition — how it works

The good news is that electronic ignition for the average modern car has boiled down to a recognisable standard formula, with a long track record of reliability. The bad news is that if it *does* go wrong, you can't fix it yourself. Having a circuit diagram is no help (which you won't be able to get hold of anyway); both the sensor and the amplifier are sealed in resin and you can't get inside without destroying them. And assuming you could get into the amplifier you will most probably find thick film resistors bonded straight onto a ceramic base which they share with other micro-mount components and a very specialised custom chip, with which you will be able to do nothing.

The history of transistorised ignition goes back as far as the 1960s. Unfortunately semiconductors of the time, being made of germanium instead of silicon, were somewhat fragile, requiring that special *beefed-up* ones be manufactured to cope. Consequently electronic ignition was expensive and usually only found attached to similarly unaffordable sports cars.

Auto electronics projects

Timing sensors

In the 1970s, solid state ignition with three versions of timing sensor proliferated. The simplest was the so called *transistor assisted ignition*, which still required a mechanical switch. The second type had an opto-electric timing sensor, which might use either visible light or an infra-red coupler. Here the beam is interrupted by a rotating shutter with blades like a fan. The third type uses a magnetic sensor.

Many of these were available as after-market *bolt-on* kits for both cars and motorcycles. After some twenty years only one type has come out on top as the simplest and most reliable — the magnetic sensor.

The sensor generates an electric pulse which triggers the amplifier, which in turn drives the coil primary. Figures 2.4(a) and (b) show the now archetypal, standard design in operation. Here a permanent magnet couples to a ferromagnetic element which is mounted on the distributor shaft and rotates with it. As this element rotates, the strength of the field varies, being largest when the air gap is smallest. The time varying magnetic field induces a current in the coil which is proportional to the rate of change of the magnetic field, and which outputs a voltage waveform as illustrated in Figure 2.4(c). Each time one of the teeth, or ridges, on the rotor passes under the coil's axis, one of the sawtooth shaped pulses is generated. The rotor has one tooth for each cylinder and the voltage pulses correspond to the spark time of the relevant cylinder. Figure 2.4(d) shows an advanced example of this idea following exactly the same principle, except that the rotor is a *star* shaped wheel and the static magnetic system has a corresponding number of poles, in this case six of each, for a six cylinder engine.

Auto advance

One reason why this triggering method has come out on top over rival designs is simply due to one staggering implication. Because the system is magnetic; it is, in effect, a very simple a.c. generator on a small scale, and its output is, therefore, proportional to the driven speed. What this means is that at slow rotor speeds the output voltage is low, while for higher speeds the output is also higher by a proportional amount. If the trigger threshold of the amplifier's input is voltage dependent, then triggering can be made to occur at the required point anywhere on the leading slope of the output waveform. Figure 2.5 shows how, from different output levels as produced by corresponding rotor speeds, the trigger level is near the peak of the slope if the output is low, and near the beginning if it is high. At a stroke, what we have here is, by way of an added bonus, an automatic ignition advance mechanism, and this with just one moving part — the rotor!

The need for ignition advance

While the fuel/air mixture in the combustion chamber burns at a constant rate, the engine as a whole however

Auto electronics projects

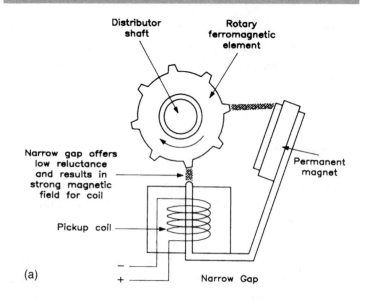

(a)

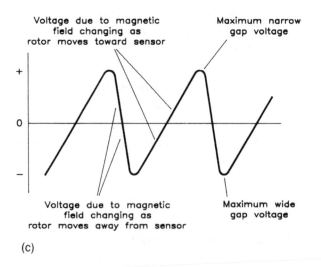

(c)

Figure 2.4   Magnetic timing sensor

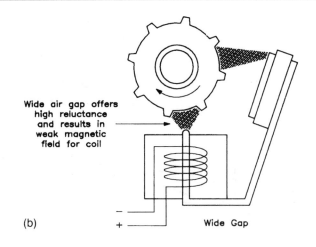

Wide air gap offers
high reluctance
and results in
weak magnetic
field for coil

(b)   −
           +   Wide Gap

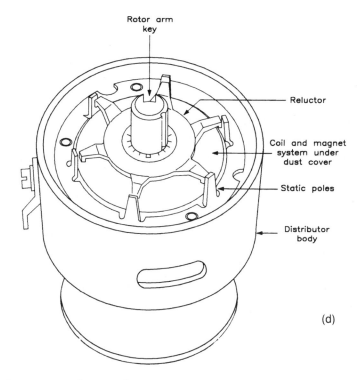

Rotor arm
key

Reluctor

Coil and magnet
system under
dust cover

Static poles

Distributor
body

(d)

Figure 2.4   Continued

49

Auto electronics projects

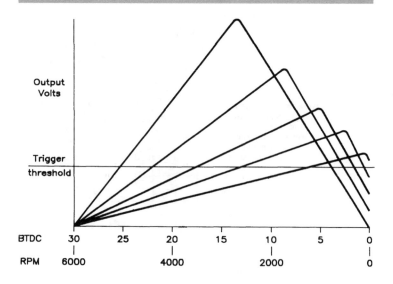

Figure 2.5   Auto-advance plot using waveform of Figure 2.4(c)

is required to operate over a range of crankshaft speeds. For this reason the moment of ignition must occur earlier at higher r.p.m. Full combustion of the fuel gas must occur during the period where the piston has full leverage on the crankshaft, and at high revs the burn actually needs to begin well in advance of this point; at lower speeds, not so much, at idle, hardly at all. The magnetic reluctance type of ignition timing sensor achieves this auto advance action in a much more linear manner than do compromised mechanical or electronic methods, and barring the odd rare mishap such as a screw coming loose, once set it does not need readjustment — for anyone who has personally endured the long drawn out process of ignition retiming, the subtleties of the operation do not need reiteration!

Furthermore, since this requirement has already been taken care of by the sensor, it makes the amplifier much simpler. Otherwise electronic advance might take the form of frequency sensitive *switches* selecting from a range of time delays, the minimum number of which is two in the crudest example of such a system. More than this requires rather more logic gates, or a microprocessor. Instead the magnetic reluctor allows the use of a comparatively very few transistors to produce an amplifier.

The electronic ignition switch

Obviously the heart of an electronic system which simulates the action of a mechanical switch to operate the coil primary in the *traditional* way is a transistor, and you might suppose that any power transistor able to carry the maximum on-time current of the primary will suffice. But oh dear me no. Remember that the primary potential is sufficient to produce an arc across the mechanical switch, and that the ignition coil as a whole, primary included, must be allowed to generate however high a voltage is necessary to bridge the plug gap? We are therefore obliged to use a high voltage power transistor, with a V_{ce} rating of several hundred volts, and such devices are notoriously inefficient, which means to say that the current gain (H_{fe}) is very small, measured in tens or less rather than hundreds.

The usual biasing method is to use a base bias resistor which typically connects directly between the transistor's base and the supply rail, and this resistor can be

Auto electronics projects

formidably beefy to provide the necessary bias current for the transistor to do its job properly, with the attendant power consumption and heat dissipation problems. I have actually seen one design where the base bias resistor is no more than 9.2 Ω!

No, that wasn't a printing error. It's an illustration of how extreme base biasing may have to be to ensure that the switching transistor achieves a saturated *on* state, essential to get the maximum available voltage across the primary of the coil and therefore the maximum primary current. Suppose, in a worst case example, that our transistor has an H_{fe} of 3 at 1 A (yes, just 3 — although fortunately later devices are better than that now), but then in order to conduct 4 A this value reduces to say <2. To ensure adequate biasing we assume a current gain of 1.5, and choose a base bias resistor with a value of 4 Ω, taking into account a base/emitter forward drop of 1 V. This resistor is then sinking 2.6 A and dissipating 28 watts; has to be removed from the rest of the amplifier to avoid cooking it to death, and be provided with its own heatsink!

Even in the case of the aforementioned design using the 9.2 Ω component, the resistor is of the high power, metal encapsulated type (see the resistors section of Maplin's catalogue for examples) and is screwed to the outside surface of the amplifier's die-cast case.

In comparison the power dissipation of the actual switching transistor is not very much at all, which seems almost perverse. This is because it performs a switching action; it is either on or off. Which leads us to the next criterion, namely ensuring that the transistor commutates (switches off) as fast as possible. This is necessary since the coil needs to be switched off quickly in order to develop its high tension output (a slowly switched ignition coil fails to make a spark).

High speed switching

Figure 2.6 shows the essentials of a typical ignition amplifier as used with a magnetic reluctance type of timing sensor. To summarise so far, TR5 is the inefficient, high voltage power transistor switch for the coil, and R9 is the base bias resistor. In this case the bias current originates from TR4, which is controlled by a Schmitt trigger comprising TR2, TR3, and resistors R3 to R6. The Schmitt trigger is essential to produce the fast edged switching waveform from the slower changing input, provided by TR1.

TR1 is the basis of the input stage which incorporates the input level threshold as indicated in Figure 2.5. This consists of diode D1 and the base/emitter junction of TR1 itself, which together will not begin to conduct until the applied level is >1.2 V. This signal is of course the ramp shaped output from the sensor coil and you can see now that while the amplitude of the ramp is variable, the input threshold is constant. D1 also blocks the negative going part of the input waveform, which is superfluous, while R1 is a current limiter to protect D1 and TR1 in the event that for example the input is accidentally connected to the supply while the power is on.

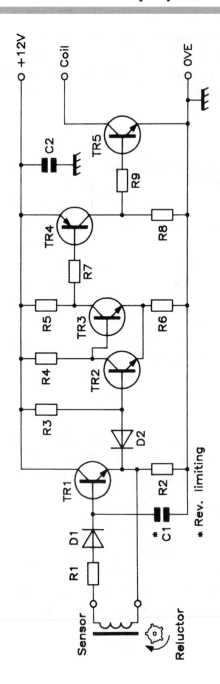

Figure 2.6 Essential ignition amplifier for a magnetic reluctor based system

Protection for the engine's mechanical bits can be provided by including C1, which acts as a *rev limiter*. While it is charged quickly via D1, this charge leaks away slowly via the base emitter of TR1 due to this device's current gain offering a relatively high impedance, and in consequence the waveform at TR1's emitter takes on a more triangular shape. As engine speed increases the mean average d.c. voltage drop across R2 also increases until a point is reached where even the lowest level of the waveform exceeds the low threshold of the Schmitt trigger; the amplifier ceases to operate and no sparks are generated.

C1 also affords some RF filtering, but it might be surprising to learn that the input leads are rarely screened. The sensor coil is of such low impedance that this is unnecessary and in any case since both these wires are run together as a pair, any externally induced current will be equally present in both, cancelling each other out.

A real working amplifier

Figure 2.7 shows a circuit which is the culmination of six months development including testing *in the field* onboard a real motor vehicle which, for earlier versions, proved to be destructive (to the circuit, not the vehicle). Such is the way of research and development, and these events made definite indications that the unit should be:

- electrically robust,
- mechanically robust; and,
- utterly weatherproof.

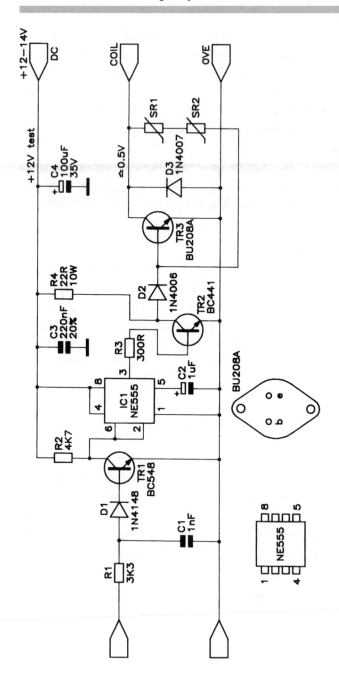

Figure 2.7 Real amplifier circuit diagram

Referring to Figure 2.7, the input stage is as described for the hypothetical amplifier earlier, with the combined diode junctions of both D1 and TR1 forming the input threshold level, and having R1 as a protective current limiter. C1 is merely an HF filter in conjunction with R1 and does not provide any rev limiting.

To reduce component count, the fast switching action needed to sharpen the pulse produced by TR1 is provided by IC1, a 555 timer IC used in an unusual way. Instead of being employed in a conventional (for the IC) manner as a monostable etc. both trigger and threshold inputs (pins 2 and 6) are tied together to exploit the behaviour of the internal bistable, forcing a Schmitt trigger action. The 555 was chosen because the output structure can source the driver stage, TR2, directly without the need for any more transistor amplifiers.

While there is no input and TR1 is off pins 2 and 6 of IC1 are high and the output pin 5 is low, so that TR2 is also off, allowing the bias resistor, R4, to saturate the main transistor switch for the ignition coil, TR3, and the coil is *on*.

Upon an input ramp voltage from the timing sensor exceeding the combined threshhold levels of D1 and TR1, TR1 conducts and quickly pulls the trigger input down to $<1/3$ of the supply level, causing IC1 to change state and switch on TR2, which clamps R4 to ground and deprives TR3 of base drive current. The coil is switched off, IC1 is reset when the ramp is completed as TR1 collector goes high again, and the system is ready to generate another spark.

Auto electronics projects

Note that all stages use the 0 V rail as the sole reference and are thus immune to supply rail fluctuations, which will occur often in the range of 12–13.8 V especially if an electro-mechanical regulator is employed, and can be less than 9 V while the starter motor is giving the battery a hard time.

Electrical safeguards

The other area of electrical weakness is concentrated on TR3. This is because of some horrible punishments that the ignition coil will try to inflict on this device. From the range of high voltage power transistors readily available the only one to prove itself electrically tough enough to be truly reliable is the long standing, TO3 packaged BU208 device designed for use in colour TV line timebases and switched-mode power supplies. The BU208A version is preferred for its lowest saturated V_{ce}, essential to ensure maximum voltage drop across the coil and reduce power dissipation in the transistor itself to a minimum — it is the more expensive version, but that can't be helped. The device has a V_{ce} rating of 700 V and a reasonable H_{fe}, which reduces bias resistor heat dissipation and power loss, as this component (R4) has a conservative value of 22 Ω(!). However TR3 still needs two essential protection schemes.

One of these must cope with ignition coil back e.m.f., which, without a power sapping condenser (see earlier) is excessive. But surely this can only occur without a spark plug as a load, else how can this happen where there is an air gap which must strike and conduct and thus limit both the coil's primary and secondary voltages? The truth is that, comparatively speaking, the air gap takes a long time to respond. Until this happens it is as if there were no load at all and the coil shoves up the potential enormously. A very simple calculation can be made to get some idea of the theoretical magnitude of back e.m.f. from a coil by:

$$\frac{\text{voltage drop across coil}}{\text{commutation time}}$$

where commutation time is the time taken for the switching device to switch off, which is of course not truly instantaneous. Assuming for example a commutation time of 100 ns which even for a BU288 is very much on the slow side, we get (in theory):

$$\frac{12\,\text{V}}{100\,\text{ns}} = 120,000\,\text{V}!$$

This is what we get on the *primary side.* In practice however it will be precisely 1,400 V. Why so? Because this is the designed collector to base (V_{cb}) limit of a BU208, never mind that this value is double the maximum V_{ce}! The base/collector junction is breaking down in the reverse direction like a Zener diode, and it is not supposed to be used in this way. Damage is cumulative and the device may fail after even some tens of hours of apparently fault free operation.

The voltage limiting protection scheme in Figure 2.7 comprises identical components SR1 and SR2, which are nothing more elaborate than two mains transient suppressers in series. This component is a Metal Oxide Varistor (MOV), the resistance of which is voltage dependent. It has a knee voltage of 340 V (that is, 1.414 x 240 V), which is the peak value of the mains supply. Up

Auto electronics projects

to this point its resistance is high, but reduces considerably as soon as its knee voltage is exceeded, and is normally used to prevent voltage spikes which would otherwise exceed the peak mains value from entering mains powered equipment.

Originally it was assumed that two of these in series would be sufficient to limit coil e.m.f. to 680 V (within the maximum V_{ce} of TR3) on their own, but in reality they are unable to cope. Consequently they have to achieve the desired objective by the alternative means of providing feedback to TR3 base and letting TR3 do the actual limiting instead. In other words, TR3 is made to switch off up to the 680 V point and then holds this until the e.m.f. value falls below this level before switching off properly. Reverse blocking diode D2 detours the current from SR2 to TR3 base so that it doesn't go straight to ground via TR2.

The other protection scheme is a provision to prevent the voltage across TR3 being reversed, i.e. <0 V, which is inevitable since the ignition coil still resonates after the spark extinguishes, for while there is no condenser there is still interwiring capacitance, together with that between TR3's case and its heatsink. The ringing is now high frequency and very short in duration, but still very much alive and kicking. This is the duty of D3.

Insulation problems

Experience has indicated that a greaseless TO3 insulator is more reliable than the traditional mica variety for heatsink mounting. If the mica is not 100% perfect then any cracks are weaknesses which can be perforated by the high voltage pulses. In the final design the unit was housed in an extruded modular alloy case (see Photo 2.1), with which a slide-in TO3 compatible heatsink was used. Although this item comes complete with screws, nuts and insulator bushes, insulator sleeves were cut from separately available TO3 bushes and pressed into the holes before mounting the entire assembly in the correct position on the stripboard ready to slide into the case, as can be seen in Photo 2.2.

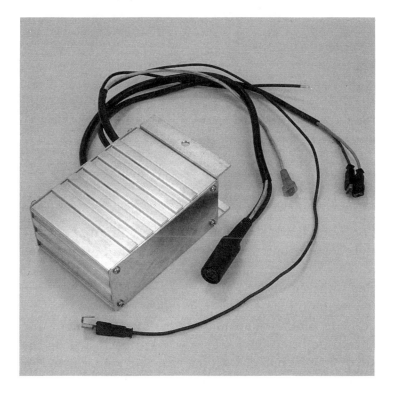

Photo 2.1   A complete home-made ignition amplifier in its case

Auto electronics projects

Photo 2.2 The stripboard assembly of the circuit of Figure 2.7 with heatsink in position and remote R4 on separate board

Mechanical considerations

Components which are at risk from vibration, e.g. upright PCB mounted electrolytics, should be supported at their base with blobs of flexible rubber sealant. IC1 was soldered directly without a socket, or else in service oxidation may cause continuity problems. R4 is a ceramic block encapsulated 10 watt component and should be fitted on a separate board such that its top surface is in contact with the case and soldered in this position during a test fitting. At final fitting this top face can be smeared with heatsink compound to fill-in the rough surface. R4 then uses the case as a heatsink.

The reason for the enormous number of external cables, evident in the example shown in Photo 2.1, is that this unit contains an identical pair of these amplifiers for a specialised motorcycle application, so there is plenty of room for one in the case!

Transistor assisted ignition

Transistor assisted ignition simply means that a *conventional* mechanical timing switch, such as a contact breaker, is not actually used to switch the coil directly but controls a solid state switch instead. The circuit of Figure 2.7 could be used in this role, by merely adding an extra 22 Ω 10 W resistor between the input and supply, as a load for the contact breaker. This will greatly increase the life of a pair of normal contact breakers, which will consequently require much less frequent timing readjustment, after which the vehicle will operate efficiently for longer periods with less damage to the environment. In addition, switching speed is faster making more energy available to the spark, although actual improvement is difficult to measure.

It is worth a mention however that the ignition coil must be a *normal spec* type with a resistance of 3–4 Ω, and *not* a high current, high energy type, these types will destroy the amplifier!

Testing

To be prudent you can check the operation of the amplifier before fitting into the vehicle. A simple test requires a 12 V power supply of up to 4 A output (or a car bat-

Auto electronics projects

tery), and a spare ignition coil. The amplifier on its own draws approximately 500 to 600 mA. By wrapping some tinned copper wire around the + terminal of the coil and looping the other end into the HT socket, a simple spark gap should be formed. This type of system must not operate without a spark gap for a load, or else it is likely to fail.

With the coil wired in, the repeated application of a 1.5 V cell to the input should produce crackingly healthy blue sparks. For the transistor assisted version, earthing the input lead for *on* and release for *off* will have the same effect. While *on,* the output (− terminal on coil) will be 0.5 to 1 V.

A more elaborate test rig is illustrated in Figure 2.8. The battery charger simulates an active charging system. The primary coil voltage can be monitored by an oscilloscope using a x10 probe for an effective sensitivity of 100 V/cm on the 10 V/cm range. It is *very important* that the probe's trimmer be precisely calibrated for an exactly flat frequency response using a high quality squarewave signal! The coil's primary winding provides a good representation of what's going on at the secondary output end, which can be seen on an 8 cm high graticule with the baseline set on the bottom or second line.

You may need to turn the brightness up and shade the screen well, as the whole event is over in less than 3 milliseconds. The trace should look like that shown in Figure 2.9.

Note that the primary's representation of the gap conduction voltage level is quite low at 80 or 90 V, but this is because the air gap is at normal atmospheric pres-

64

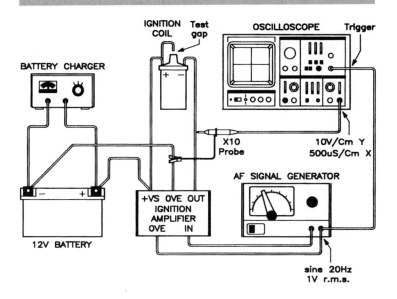

Figure 2.8 Test rig for monitoring amplifier output at scope

sure. While providing sparks for a real engine this level actually wanders about all over the place in direct proportion to the gas density in the combustion chamber, being at its greatest while this is high during acceleration, and lowest during the over-run while the throttle is closed. It is for this reason that the upper limit is designed at 680 V and the BU208 chosen in order to provide plenty of *headroom*: a different output stage with a lower voltage transistor *will not work properly* (as it stands, the design has been found to handle compression ratios of >10:1). This behaviour also explains why any insulation weakness always breaks down during acceleration. Such a breakdown is usually total, as I found out the hard way, leaving me stranded. So take note!

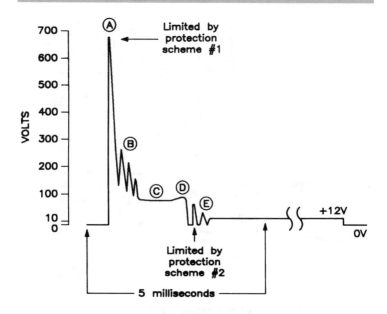

Figure 2.9 Oscillograph produced by test rig: (a) initial e.m.f. pulse: (b) spark gap ionisation time; (c) gap conduction time: (d) gap extinguishing moment: (e) ringing period

C.D.I.

Who remembers D.I.Y. clip-on ignition boosters. At one point during the late 70s, the popular motor accessory shops were crawling alive with these things. The selling point was the third principle mentioned earlier — capacitive discharge ignition.

CDI employs the ignition coil in a totally different manner, namely as a form of pulse transformer. The advantage is that the coil is no longer an appreciable part of the electrical load as in a more conventional switched system; it does not have a heavy current flowing in it for a large part of the time and consequently has an easier life promoting reliability. In addition, overall power consumption for the ignition system as a whole is much lower and is in fact proportional to engine speed. As well as by the much reduced power requirement, cold winter starting is aided by the very high energy spark that CDI can generate, which, if the designer is careful, is still available even if the battery voltage is very low during starting.

CDI is electrically efficient like no other alternative system, producing enormous sparks for a miserly few hundred milliamps of supply current. Past experiments by this author with home grown CDI designs have produced sparks of $1^{1}/_{2}$ inches! Figure 2.10 shows a typical system in block form, and individual designs do not deviate much from this.

The heart of the system is a d.c.–d.c. converter, which produces a high voltage first (as opposed to the switched method which derives it at spark time by switching the coil off) directly from the low tension supply. It is stored by capacitor C1 which is in series with the coil primary winding.

The input stage receives a signal from a magnetic or other form of timing sensor or a contact breaker, and trips a pulse generator, usually a monostable. The output pulse triggers on CSR1, which clamps C1's *live end* to ground. The coil primary suddenly finds something in the region of 500 V across it, and commences discharging C1. In the process, the discharge current induces a current in the secondary winding, where the primary voltage is multiplied by the turns ratio, producing a spark at the HT output. The counter-e.m.f. from the coil primary that follows turns CSR1 off again. While all this is going on,

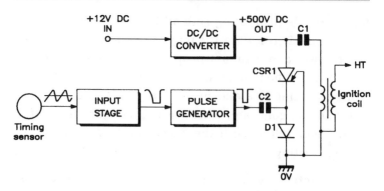

Figure 2.10   Capacitive discharge ignition block schematic of essential parts

the converter's output is effectively short-circuited to earth, and it must be designed in such a way that it is not damaged by this.

The system is that simple, and easy to design, but latterly is by and large *not taken seriously* by most motor manufacturers. Why should this be? Because of two inherent, unavoidable flaws in the principle.

One of these is to do with spark conduction time. The truth is that this depends on capacitor discharge time, and as a result can be appreciably shorter than that of a conventionally switched coil. This means less gap conduction time in the combustion chamber and, to be blunt, less than ideal ignition of the fuel gas. In reality a better burn (and less waste and pollutants) results from a medium energy long spark than a high energy short one — although this also depends on how the combustion chamber design can make the best use of it; with some older shapes, which are so inefficient in the first place, it won't make much difference.

The obvious answer is to increase the value of C1 to increase conduction time, but this aggravates the second problem — which is that the capacitor should be completely recharged prior to the next spark moment. Suppose that C1 were increased to 1 μF to provide a 4-cylinder engine with reasonable sparks up to its peak output speed of 6,000 r.p.m. This requires 200 sparks per second, further requiring C1 to be recharged in the space of <5 ms. This needs a charging current of 100 mA, which can be proved by:

$$\frac{100,000\,\mu A}{1\mu F} \times 5\,ms = 500\,V$$

and the average power consumption of the converter increases, by coincidence, to 50 watts — I say by coincidence because this is also the average for a conventionally used ignition coil. In practice the spark strength of CDI always drops off along a steadily worsening curve at higher r.p.m., aggravating incomplete combustion, already compromised by gas flow problems and such. This is not to say that switched ignition doesn't have a similar behaviour, but the roll-off of a switched coil is less acute, and in any case it is easier to select or manufacture the coil for the job required.

To be fair though, CDI is not a totally duff idea, but, should you be toying with the idea of investigating the principle yourself, be advised that, in order to be able to deliver the required goods with any semblance of real usefulness, the converter should follow a high frequency type of switched mode power supply principle, using a ferrite cored transformer, and *not* use a mains transformer *in reverse*! Mains transformers are designed to tap power from the mains at mains frequency, and are not very good at doing anything else. Given the *short-circuited* output problem, the converter could be a single-ended flyback converter design.

The future

One possible forthcoming innovation for cars is distributorless ignition. Instead of a mechanical rotor delivering the HT current to the required plug as necessary, one iteration of the principle is to use high voltage rectifiers in a floating secondary circuit to steer HT to the desired pair of cylinders in a 4-cylinder engine, the other cylinder, which does not need a spark, is on its exhaust stroke and so a spark here is known as a *wasted spark*. The ignition coil primary is double-ended and operated in *push-pull* mode by a pair of switching transistors; the direction of the secondary pulse determines which pair of plugs will receive the current via the diode matrix, and the transistors will no doubt be under the control of an engine management computer.

A variation will use two ignition coils, also with floating *open-ended* secondary windings but terminated straight to a spark plug at each end. Again the relevant pair of pistons move together but their valve timing is 180° out of phase, so that while one is on its compression stroke, the other is on its exhaust stroke.

In actual fact motorcycles have featured duplicated complete ignition systems, and the *wasted spark* technique for many years, and it is only a question of time before motor cars follow suit and become equally distributorless.

3 Microcontrollers

The microcontroller is *the workhorse of the modern electronics industry*. That statement may be strong, but it is not an exaggeration, for it is becoming increasingly difficult to purchase any significant piece of electronic hardware that does not contain one or more of these complex ICs.

A microcontroller (μC), otherwise known as a single chip microcomputer unit or MCU, is effectively a complete computer control system integrated onto a single chip of silicon. Referring to Figure 3.1 the main functional blocks of the microcontroller are:

● microprocessor core: with optimised instruction set for real time control,

Auto electronics projects

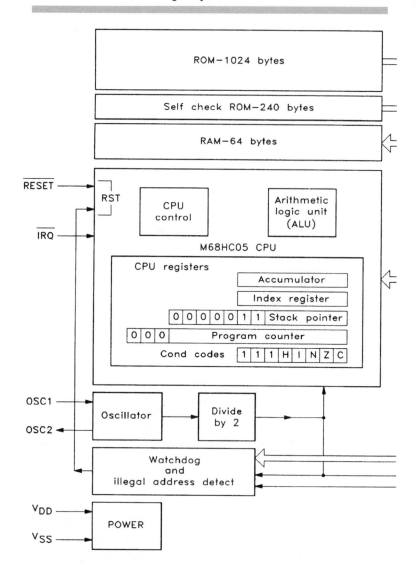

Figure 3.1   MC68HC05J1 MCU block diagram, showing the
basic functional blocks common to all microcontrollers

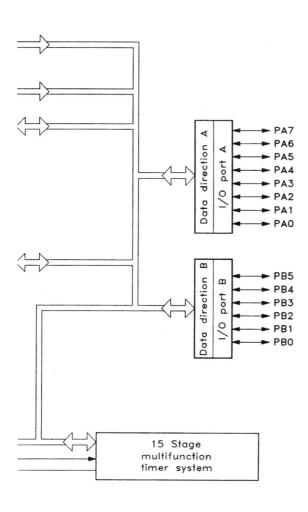

Figure 3.1   Continued

Auto electronics projects

- memory; usually ROM to contain the control program plus RAM to hold variables during program execution,
- I/O and on-chip peripherals; these allow the MCU to communicate with the hardware of the real world application that it is controlling. These peripherals range from simple digital input/output (I/O) ports to complex analogue-to-digital (A-to-D) and digital-to-analogue (D-to-A) converters and timer systems. Table 3.1 lists some of the *peripherals* that are available on current microcontroller families.

Microcontrollers are available in a range of complexities and power (and therefore price), making them suitable for a very wide range of applications where they can replace standard logic or more complex microprocessor based solutions. The advantages of the MCU over these traditional solutions are, reduced chip count, which brings cost; reliability and size bonuses; and greater flexibility for the designer — allowing easy modifications to the functionality of the application via the software. These advantages coupled with the devices' relatively low cost (typically from £0.75 in high volume) have led to microcontrollers being used in a great breadth of applications. With a few exceptions such as industrial control, these MCU applications can be split into two groups; automotive and consumer.

Table 3.2 gives a non-exhaustive list of microcontroller applications in these two areas. The intention of this chapter is to give the reader some more insight into a few of the automotive applications that depend on microcontrollers, and to highlight the properties of particular MCUs that make them suitable for each discussed application.

MCU peripheral	Function
Digital I/O port	The basic hardware used by the CPU to access the outside world (read switches, drive LEDs, etc.).
Timer	One off the most common and useful MCU peripherals — allows timing tasks to be accomplished while the CPU does something else.
Serial port	Both synchronous and asynchronous ports are available allowing fast serial communications over short or long distances respectively.
VFD port	Special high voltage output port for driving vacuum fluorescent displays.
LCD port	Special low voltage output port for driving LCD displays. Usually includes multiplexing for large displays.
A-to-D	Analogue-to-digital converter used to read a variety of sensors. etc.
PWM or D-to-A	A pulse width modulated output that can be filtered to produce a programmable analogue voltage, thus acting as a digital-to-analogue converter.
Watchdog timer	A special type of timer that guards against CPU errors and resulting software runaway.
EEPROM — in addition to ROM	Re-programmable memory that can be used for calibration purposes or for a non-volatile data store.
PLL	Phase locked loop. Used in tuner applications such as TV and radio.
RTC	Real time clock. Special timer designed to count in real time, i.e. seconds, minutes and hours.
Wake-up port	Modified digital I/O port that can generate CPU interrupts when an input signal changes.
DTMF	Dual-tone multi-frequency generator, used in *tone dialling* telephone applications.
OSD	On screen display. A character generator for showing messages on a TV screen.

Table 3.1 Commonly available on-chip microcontroller peripherals

Auto electronics projects

The automotive industry is widely recognised by semi-conductor manufacturers as being the performance driver of the microcontroller market. Originally using microcontrollers with 4 and 8-bit buses, the automotive designer's quest for more processing power for some applications, such its engine management, has pushed the semiconductor industry into designing first 16-bit and now 32-bit MCUs. Some cars being designed today have more processing power under the bonnet than an average PC!

A well recognised trend in the automotive industry is to introduce new features on up-market cars and then migrate them down onto their mass market vehicles as reliability and user acceptance are proven, and costs come down. This explains why many of the features available on today's cars (such as electric windows) were yesterday only available on expensive luxury models. However, in many cases these systems are using yesterday's *dumb* technology and many of the microcontroller applications of Table 3.2 are still the domain of up-market vehicles. As the technology migration trend and *green* legislation continue, this situation will change and within a few years all cars will contain more microcontrollers than wheels! See Figure 3.2.

Interfacing MCUs in the automotive environment

There is a fundamental problem with using micro-controllers, or digital logic in general, in an automobile; the vehicle electrical system is invariably 12 V and logical devices work at around 5 V, and would be severely

Automotive	Consumer
Engine management	Television
Alarm system	Microwave oven
Anti lock braking	Telephone
Central locking	Video cassette recorder
Trip computer	Washing machine
Dashboard	Remote control system
Electric windows	Toys
In-car entertainment	Fridges and freezers
Active suspension	Alarm system
Multiplexed wiring	Radio
Seat adjustment	Compact disc player
Electric mirrors	Satellite receiver

Table 3.2 Typical microcontroller applications

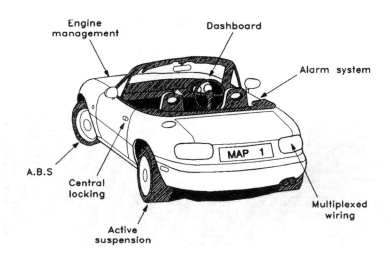

Figure 3.2   Soon an average car will contain more microcontrollers than wheels!

damaged if connected directly to a 12 V system. This means that a supply for the MCU must be derived from the 12 V supply using a regulator circuit, and that all inputs to the device must be buffered from the 12 V world around it. The MCU is also incapable of directly driving automotive loads, so that external drive circuits must be employed to interface the logic outputs to the 12 V loads. The situation is actually even worse than this initial statement implies; the automotive environment is one of the harshest known, with extremes of temperature and the system voltage varying considerably depending on the condition of the battery and whether the vehicle engine is being cranked (when the voltage drops considerably). The biggest problem however, is the ignition circuit. When the ignition coil switches, large voltage impulses (50 to 100 V) can be generated on both rails of the entire electrical system. Although of short duration, these pulses would spell disaster for a logic circuit input. For this reason great care must be taken when designing protection circuits for the electronic hardware in cars. Despite these problems and the associated costs to counter them, the outlay is justified due to the benefits brought by electronics and microcontrollers, in particular to the automobile. In the following discussions and examples, the protection and drive circuits may not always be shown for simplicity, but the reader should be aware that these precautions have to be taken in all automotive microcontroller applications.

Electric windows

This is one of the most common *electrical goodies* to be fitted to many cars. Figure 3.3 shows the traditional dumb

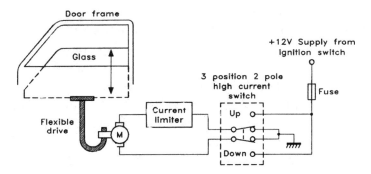

Figure 3.3   Conventional electric window circuit (duplicated for other doors)

electric window circuit that is in common use today. The switches directly control the supply current to the motors, thus propelling the window in the desired direction. When the window reaches the end of its travel there is no cut out, instead the motor simply stalls and the current is limited to a value that does not damage the motor windings. You can observe this by trying to raise both closed windows in a car when the engine is idling the engine r.p.m. will drop appreciably due to the heavy loading on the alternator. Although this system works quite well, it does have a couple of problems. The first of these is quite a major safety concern and stems from the fact that to deal with icy windows or a dirty mechanism a powerful motor is deployed. The problem is that if an obstruction is placed in the way of a closing window the motor will exert a great deal of force before it stalls; that obstruction could be a child's neck. The second problem is more of an annoyance than a real problem and it concerns the amount of time that the driver must keep his finger on a small button to fully open or close the window.

Auto electronics projects

Both these problems are solved by the *intelligent* MCU based system, shown in Figure 3.4. Here the switches and sensors are connected to inputs of the MCU and it in turn controls the motors via output ports that switch external drivers. The sensors inform the microcontroller that the window has reached the end of its travel and the MCU can stop the motors. This positional feedback along with the current sense means that the MCU can immediately detect when an obstruction other than the end-stop has caused the motor to slow or stall instead. In these cases the MCU can now take evasive action by stopping and reversing the direction of the window for a couple of inches thus releasing the obstruction. The MCU also allows the option of one-touch open or close, either via an additional button, or by counting how long the normal button is held for — e.g. if the button is pressed for more than 2 seconds then the MCU assumes a full motion of the window is required. Although these features could be implemented using logic control, the integration and very low cost of a simple MCU such as the MC68HC05J1 from Motorola make it the ideal choice. This device is supplied in a small 20-pin package and has only 1 K of ROM onboard to store the program, along with the CPU and a simple timer (Figure 3.1). However, these limited features linked with low cost make it the ideal device for displacing clumsy logic solutions.

Central locking

That great innovation for the wet British climate, central locking, has traditionally been operated via a switch in the lock mechanism of the front doors, but in recent years a new development has made this feature even

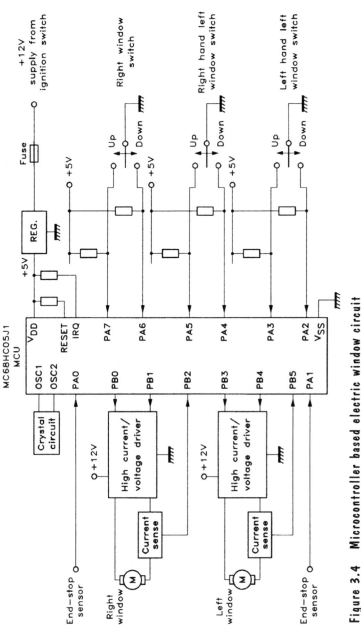

Figure 3.4 Microcontroller based electric window circuit

81

Auto electronics projects

more desirable — remote central locking. In this set-up a remote key uses a transmission by radio, or more commonly infra-red (IR), to activate the central locking from a wide angle and considerable distance from the vehicle — Figure 3.5 shows the schematic of such a system. The transmitter uses either a very basic microcontroller or, more commonly, a dedicated logic device such as the MC145026 IC. Instead of using a keypad to determine which code to transmit, the device has its inputs fixed in the factory, into a certain combination of logic levels, so that it will always transmit the same code. The number of inputs allow a large number of different codes to be configured — just like the number of levers in a padlock.

Although matched pairs of transmitters/receivers could be employed in this application, the logistics of keeping track of which *key* belongs to which car during production are obviously difficult, never mind how you would handle an owner losing his key and requesting a replacement! For these reasons, intelligence is employed in the receiver to allow it to be customised after production. The microcontroller chosen for the job will include some on board programmable non-volatile memory (EPROM or EEPROM) that can be used to store the codes of matching transmitters. This customising of the receiver is often performed by the dealer, just before the new owner gets his car. The memory size of the MCU allows for several key codes, so that multiple keys can be used by different family members. Secure software can be employed to prevent someone from trying to cycle through all the valid codes for the transmitter type until the correct one is found. In its simplest form this could just involve ignoring incoming IR codes, for a couple of seconds, after an invalid code has been received — with so many codes

82

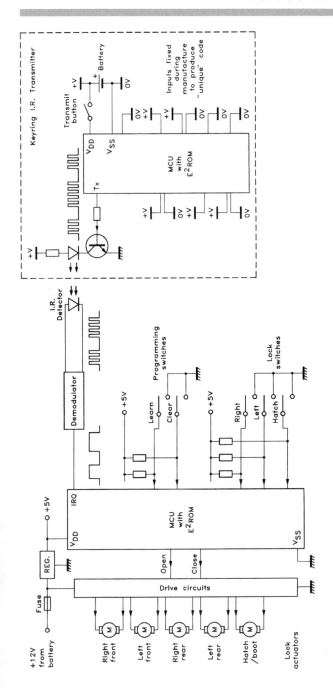

Figure 3.5 Intelligent remote central locking system

to cycle through, this would make the job overly time consuming for the potential intruder.

Since the receiver must remain powered up at all times, low power consumption is of vital importance. For this reason the MCU will invariably be a CMOS device, with a special low power *sleep* or *stop* mode, where the power consumption will be in the order of microamps. Any incoming signal will wake the MCU, via the interrupt pin, and it will receive the code and operate the locking mechanism (either solenoid or motor driven), if it matches one of the valid codes stored in its memory. A suitable device for this application would be the MC68HC05P8, which is a close family member to the previously discussed J1 device. Its distinguishing feature for remote central locking is the 32 bytes of onboard EEPROM that can be used to store several transmitter key codes.

Engine management

Engine management in this context means having complete control over an engine's ignition timing and fuel mixture on a cycle-by-cycle basis. The trend in increasing engine management performance has been driven by the tightening of emissions regulations around the world. This is the real performance end of the microcontroller market, and it has been responsible for the growth in complexity of the µC's on-chip timer systems for, as we will see, engine management involves a lot of time-critical tasks. Before discussing where microcontrollers fit into this application, a brief explanation of what is involved in engine management and how it has been tackled in the past would be beneficial.

Figure 3.6 shows the four stages of a complete cycle of a four-stroke internal combustion engine. In the first stroke, the piston is travelling downwards with the inlet valve open, thus drawing in the air/fuel mixture from the inlet manifold. In the second stroke, the piston rises with both valves closed, thereby compressing the mixture. As the piston reaches the top of its travel (top dead centre or t.d.c.), the spark plug is fired to ignite the mixture. The third stroke is the combustion/power stroke, when the cylinder delivers its power; the rapidly combusting mixture becomes very hot and the resulting rapid increase in pressure drives the piston down the cylinder. In the final stroke, the piston travels upwards again, with the exhaust valve open, thus expelling the remaining burnt gases. The piston is then ready to start its next downward intake stroke, and so initiate another four stroke cycle.

The problem for the automotive designer is that to maximise the power and fuel consumption of an engine (while minimising its pollutants), the timing of the ignition spark and the ratio of the air/fuel mixture must vary according

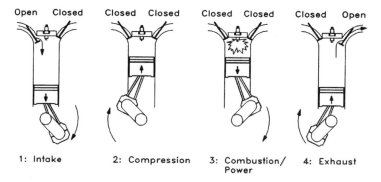

Figure 3.6   The four strokes of the internal combustion engine

to a number of factors. The most significant of these factors are engine speed, temperature and engine load. The job of engine management is to control the ignition and fuelling of the engine, keeping it as close to its ideal operating conditions as possible.

Ignition

On average it takes 2 ms to complete the combustion process after the ignition spark has been fired. Since the aim is to have maximum pressure in the cylinder just after the piston has passed the top of its stroke (too early and the pressure would inhibit the pistons upward travel; too late and power is wasted in the downward stroke), it is necessary to fire the spark before the piston reaches t.d.c. It is customary to represent this *ignition point* as the number of engine degrees before top dead centre (b.t.d.c.). As the piston travels faster and faster with increasing engine speed, and because the combustion process takes the same length of time, a fixed firing angle for the spark would result in maximum pressure occurring further and further into the downward stroke, so wasting power and increasing fuel consumption. For this reason the ignition point must be advanced (more degrees b.t.d.c.) with increasing engine speed. Traditionally this has been accomplished with the centrifugal advance mechanism in the distributor.

Another factor which influences the ignition timing is the engine load, which can be shown to be proportional to the amount of air inducted by the engine. Historically this factor was taken into account by connecting a pipe from the inlet manifold to the distributor advance mechanism (the vacuum advance). This mechanical system (which has remained virtually unchanged for many years) is reliable, but only allows crude control of the ignition timing, resulting in compromises in the engine performance and great difficulty in reaching today's emission regulations.

Mixture control

For an engine to run well, a specific air/fuel ratio must be maintained. The theoretical ratio of fuel to air for complete combustion (and therefore maximum economy and lowest emissions) is just under 1:15 in weight (or alternatively 1 L of fuel for every 10,000 L of air in volume). In practice, maximum fuel economy is obtained with around 20% excess air, while maximum power is obtained with approximately 10% air shortage. Since engines normally run at part-load, the fuel system is designed for maximum economy at this point and the mixture will be richer at idle and maximum power. The task of the carburettor, or fuel injection system, is to produce the best mixture for the engine under its current operating conditions. The simple mechanical carburettor again compromises the fuel mixture under different conditions and the trend today is towards electronic fuel injection, where a precise amount of fuel can be delivered to the individual cylinders.

Because ignition timing and fuel mixture are both dependent on the same variables (engine speed, load and temperature), it makes a lot of sense to combine the control of both into a single unit — the so-called electronic engine management system. With its ability to read

Auto electronics projects

sensors, perform high-speed calculations and measure time, the microcontroller is the ideal device for engine management.

Figure 3.7 shows a block diagram of a typical system based on a high-performance microcontroller. Engine speed and angle are both obtained from a single inductive sensor that generates electrical pulses when teeth on the flywheel pass by. To provide a reference point for determining the engine angle, one or more teeth are omitted from the flywheel, thus producing a pulse period twice or more than the normal. Alternatively, but less common, an extra tooth may be present resulting in two pulses each of half the normal period. This means that to determine engine speed and angular position, the microcontroller must perform two basic tasks:

● it must detect the missing/extra tooth and then count teeth to determine the engine angle,

● it must track the time between adjacent teeth, and from this calculate the current engine speed.

As there are typically 30 to 60 teeth on the engine flywheel and a typical engine has to be designed for an 8000 r.p.m. maximum, the pulse period from the flywheel sensor can be less than 125 µs. Clearly then, if the µC used software loops to count the periods of the incoming pulses, there would be very little processing time left to use the data obtained, even if it didn't have other signals to measure as well. For this reason independent timer systems on board the microcontroller have evolved to lessen the load on the CPU. These timers use an input *capture* mechanism to *time tag* incoming pulse edges against a free running counter timebase, and then inter-

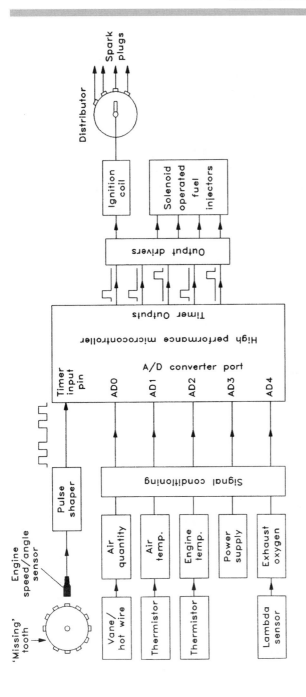

Figure 3.7 Simplified block diagram of an engine management system

rupt the CPU to tell it to read the captured time. The following section illustrates how the timer system interacts with the CPU on Motorola's M68HC11 microcontroller in order to determine engine speed and angle.

The diagram in Figure 3.8 shows a simplified block diagram of the Timer and Pulse Accumulator systems onboard the Motorola M68HC11 MCU, a particularly popular device for current engine management solutions. The heart of the timer system is a 16-bit free-running counter that the CPU can read via two 8-bit registers, TCNTHI and TCNTLO. The main purpose of the counter is to act as a timebase for the *input capture* and *output compare* functions. The input capture function allows a transition on an external pin to be *timestamped* by latching the value of the free-running counter at the time of the transition. The CPU can then read the latch at a later time and get an exact record of when the transition occurred.

The output compare, or match, function is the inverse of the input capture; it allows the CPU to schedule a change in the state of an output pin, at a precise time in the future, by writing a value into the 16-bit compare register. When that value is matched by the incrementing free-running counter, the output will change state. The M68HC11 has various combinations of input capture and output compare pins available on several of its family members. The pulse accumulator is an 8-bit counter that is clocked by a specified transition on an external input pin. The CPU can write any value into the counter, and can read it at any time. The pulse accumulator can generate an interrupt to the CPU when it overflows.

The timer module and pulse accumulator can be used in a number of ways to determine engine speed and angle, and to generate the necessary output pulses. The following is one such method.

The conditioned signal from the flywheel sensor is connected to both the pulse accumulator input pin and input capture pin. Both the input capture pin and the pulse accumulator pin are configured to detect a rising edge, and the input capture interrupt is enabled so that the CPU will be interrupted on every pulse rising edge. The interrupt software routine will read the captured value of the free-running timer, store it and then subtract the last captured value, to obtain the tooth pulse period in timer counts. Since the number of teeth are known, the engine speed can easily be derived from the pulse period. Since the period of every pulse is measured, the interrupt software can identify the longer period associated with the missing tooth angular reference. At this point it can clear the pulse accumulator, which will then start counting pulses (teeth). Since each tooth corresponds to a number of engine degrees, the value of the pulse accumulator is a representation of the engine angle.

To generate one of the output pulses (e.g. for an injector), the input capture interrupt software can check the pulse accumulator against the desired tooth count (minus 1 or 2, to allow for interrupt latencies). When the pulse accumulator matches this value, the CPU can schedule the start edge of the pulse by reading the free-running counter, adding an offset and writing the resultant value into the output compare register of the desired pin. The offset is a value in timer counts that corresponds to a number of engine degrees at the cur-

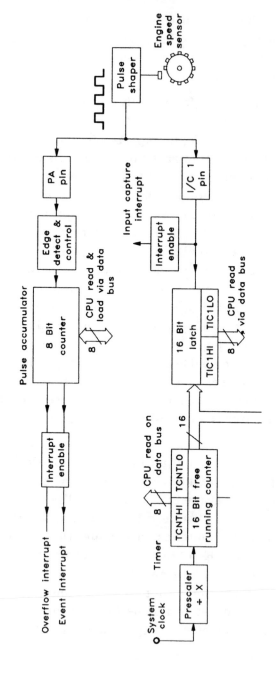

Figure 3.8   Simplified functional block diagram of the MC68HC11 time/pulse accumulator system

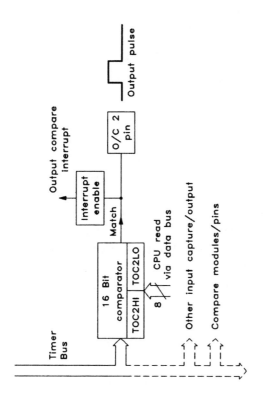

Figure 3.8 Continued

Auto electronics projects

rent speed. This number of engine degrees is the difference between the angle matched to the pulse accumulator value, and the exact number of engine degrees at which the pulse must begin — see Figure 3.9.

The other input parameters of Figure 3.7 are measured using an analogue-to-digital (A-to-D) converter, which is usually integrated on-chip as part of the microcontroller. As previously mentioned, the air inducted by the engine can be used as a measure of the engine load. A value for this is obtained, via a vane device in the air intake that operates a potentiometer, or alternatively via a hot-wire sensor. The lambda sensor is a fairly recent addition to engine management systems prompted by increasing anti-pollution regulations that have led to the use of catalytic converters. The catalytic converter is a delicate object and very stringent control of the engine emissions must be obtained if the catalytic converter is to operate

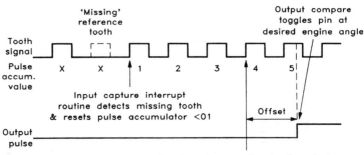

In the above example an output pulse must be generated starting at an engine angle corresponding to 5 teeth plus a bit (as each tooth is a number of degrees) from the reference mark

Figure 3.9 An example of output pulse timing

efficiently. The lambda sensor is basically a hot platinum/ceramic device that produces an output voltage which varies, depending on the oxygen content of the gas it is surrounded by. By inserting such a sensor into the exhaust manifold, it is possible to determine the air/fuel composition currently being burned in an engine. This effectively transforms the engine management system, from an open-loop control system into a closed-loop one, where deficiencies in the desired output (correct air/fuel mixture) can be detected and the input variables (ignition timing/fuel quantity) adjusted to compensate. This means that much closer control of the exhaust emissions can be maintained, helping to maximise the efficiency of the catalytic converter mounted downstream in the exhaust system.

Having measured all these parameters, the microcontroller must determine the corresponding outputs — i.e. the timing of the spark ignition pulses, and the timing/duration of the pulses which fire the fuel injectors. This is achieved by accessing the so-called *engine maps* that are stored in the memory of the microcontroller. These maps are, in fact, tables of data that hold the ignition and fuelling characteristics of a particular engine type against a number of input variables. Because it is impractical to try and store all the possible combinations of output timing versus input characteristics, a number of points are held in the map table, and the μC must then perform an arithmetic calculation to interpolate between the two closest points given, to the exact input conditions obtained from the various sensors.

As there are a number of variables to be taken into consideration, these interpolation calculations are complex and require a lot of processing power to be completed quickly, in time to set up the output timings for the next engine cycle. This is the reason why 16 and now 32-bit microcontrollers are replacing older 8-bit systems for engine management. They allow more complex calculations to be completed quickly so that closer control can be maintained on a cycle-by-cycle basis.

When the microcontroller has obtained the desired output timings, it must actually generate the pulses to fire the spark plugs and injectors. This is done via the output *match* facility of the timer system, where the CPU writes a value into a special register. When the value of the incrementing timer-counter reaches the same value as that in the register, the hardware of the timer system automatically changes the output pin state to a desired level. This mechanism allows very accurate placement of the various pulses required in the engine cycle, as we have seen from the description of the Motorola M68HC11.

The method described above, using the input capture and output match timer functions, is used in virtually all of today's production engine management systems. However, this system is not perfect as the CPU still has to respond to a large number of interrupts generated by the timer, thus slowing down its control calculations. This *interrupt overhead* has set the performance limits of today's systems, and so a new approach will be required for the even more complex control algorithms required for tomorrow's emission regulations.

Motorola has been the first microcontroller manufacturer to address this problem by introducing the innovative MC68332 device. Not only does this device have a powerful 32-bit 68000-based CPU, but is unlike any other microcontroller in that it also has a second on-board CPU dedicated to controlling timer functions. This Time Processing Unit, or TPU, is in effect a microcontroller within a microcontroller! The TPU is used to handle almost all of the interrupts associated with the timer channels, thus freeing the main CPU to spend more time on complex control calculations. At suitable points in the control cycle, the main CPU obtains new input readings from the TPU and presents new data for the TPU to calculate and schedule the output pulse timings.

Vehicle alarms

The huge increase in car-related crimes in the 1980/90s has been paralleled by an equally large increase in the demand for car alarms. Originally based on simple logic circuits and triggered directly from interior light switches, the complexity of alarms has grown to try and match the skill of the potential intruder. Figure 3.10 shows the schematic of a typical sophisticated MCU-based alarm system. Using a microcontroller in this application provides a great deal of sophistication within a very low component count, allowing the alarm to be small and thus easily concealed.

An MCU chosen for this job should have a *low power* mode since the alarm must be powered up for long periods of time without the engine running. It should also be possible to *wake* the device from this mode via several sources, so that a number of circuits can trigger the device into sounding the alarm. A simple 8 or 16-bit on-chip timer is also desirable to time the output audio/visual warning pulses, and to reset the alarm after it has

Auto electronics projects

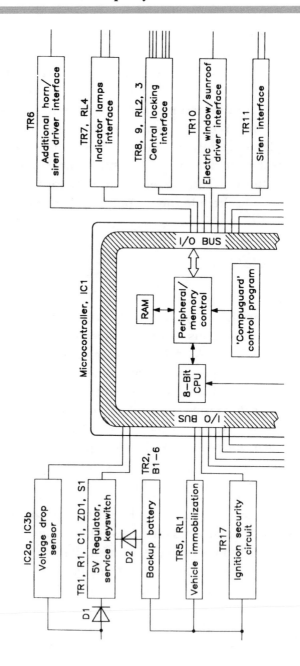

Figure 3.10 A μC-based car alarm system

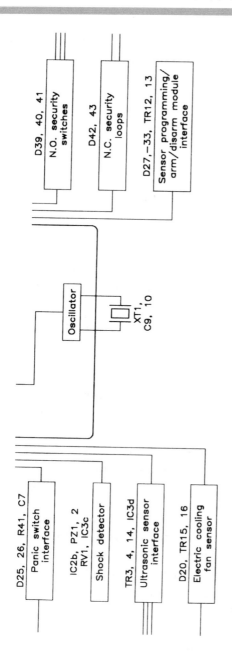

Figure 3.10 *Continued*

Auto electronics projects

sounded for a set time — this is a legal requirement. The timer can also be used to arm the alarm after a defined period, if it is not armed via a remote control.

A.B.S.

The increased performance of everyday cars, along with their increasing numbers (and therefore greater density on the roads), has resulted in a continual improvement in braking performance. This trend has included the progression from all-drum braking, drum/disc braking and ventilated disc/drums, through to the all-disc braking systems found on today's higher performance cars. The most recent improvement has been the introduction of ABS.

The Antilock Brake System does not itself increase the braking capacity of the vehicle, but improves safety by maintaining optimum braking effort under all conditions. It does this by preventing the vehicle wheels from locking, due to over-application of the brakes, and thus maintains steerability and reduces stopping distances when braking on difficult surfaces such as ice.

ABS allows shorter stopping distances than with locked wheels, due to the friction or mu-slip characteristic of the tyre-to-road interface; as a wheel brakes, it slips relative to the road surface producing a friction force. A typical mu-slip curve is depicted in Figure 3.11. This shows that peak friction occurs at about 10 to 20% slip, and then falls to approximately 30% of this value at 100% slip (locked wheel).

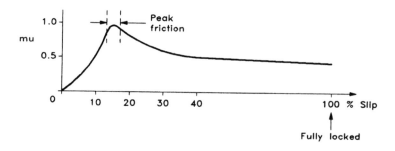

Figure 3.11   A typical mu-slip characteristic for the tyre-to-road interface

The aim of the ABS system is to control the braking force so as to stop the slip for any wheel exceeding this optimum value by more than an acceptable window.

At the heart of all ABS systems (except the all-mechanical system implemented by Lucas) is an electronic control unit (ECU) based around a powerful microcontroller. Figure 3.12 shows a block diagram of such a system. The solenoid valves that form part of the hydraulic modulator allow control of the pressure available to the individual wheel brake cylinders, independent of the force supplied by the driver via the brake pedal. These three-way valves can connect the brake cylinders to:

● the normal master cylinder circuit, so that the braking pressure will be directly controlled by the driver,

● the return pump and accumulator in the hydraulic modulator, so that the pressure in the brake cylinders will fall as the fluid returns to the master cylinder,

101

Auto electronics projects

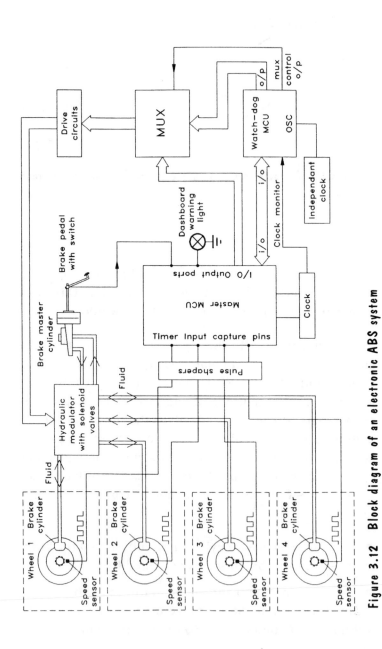

Figure 3.12 Block diagram of an electronic ABS system

102

● neither of the above two circuits, thus isolating the brake cylinder so that the pressure will be maintained at the value immediately preceding the move to this position.

The control for these valves is supplied via drive circuits from the output ports of the microcontroller.

The basis for all electronic ABS systems is the microcontroller's ability to determine the speeds of the individual wheels (although some front-wheel drive vehicles share a common speed sensor for both rear wheels). It does this via an inductive sensor and toothed ring that produce an output waveform, the frequency of which represents the speed of the wheel. This arrangement is almost identical to the engine speed sensor discussed earlier, except that since no angular position information is required there are no missing or extra teeth. It follows from this that the explanation previously given on determining engine speed also applies to determining wheel speeds in an ABS system.

In this case, there are around 50 to 100 teeth on the encoder ring, and this could result in a pulse frequency of 6000 Hz when the vehicle is travelling well in excess of 100 mph. As there can be a speed sensor on each of the 4 wheels, a total of 24,000 pulse edges have to be resolved every second. The solenoid valves in an ABS system typically have a response time of 10 to 20 ms, and the microcontroller must be able to sample the inputs at least twice that often, to resolve lock-ups in 5 to 10 ms.

Put another way, the microcontroller must be able to determine 4 independent wheel speeds from 6000 Hz signals within a 5 ms window, and still have time to carry out processing on this data to determine the new valve states. These stringent timing requirements mean that ABS systems are the domain of high performance 16-bit microcontrollers that can respond quickly to interrupts from the timer system which is capturing the speed sensor edges.

So far it has been stated that the microcontroller in an ABS system must prevent the wheel-slip value from exceeding the optimum, and we have discussed how the μC measures the wheel speeds (angular velocity). However, it may not be clear how these wheel speeds are related to the slip values that the system is attempting to control. The slip of any wheel can be defined as the difference between the angular velocity of the slipping and non-slipping wheels, divided by the angular velocity of the non-slipping wheel. This makes sense and sounds quite simple, but for one problem; how to find a non-slipping wheel? The ABS algorithm searches for the fastest spinning wheel and uses this as the reference for calculating the slip values of the other wheels. If the slip value of a wheel is greater than the peak friction value by a certain margin (i.e. the wheel is heading towards a locked condition), then an ABS control cycle is executed on that wheel.

First the microcontroller will isolate the wheel brake cylinder from the brake circuit to prevent further pressure increase. It will then recheck the slip and acceleration values to determine if the wheel is still decelerating, and whether the slip value is still exceeding the desired value. If so, then the valve position is moved momentarily to the return position, reducing the braking effort on that wheel. This pulsed release of pressure is continued until the microcontroller detects that the wheel acceleration is positive, at which point it stops reducing the pressure, and reconnects the wheel cylinder to the brake circuit to prevent overshoot of the acceleration. This entire control cycle of holding/reducing/increasing brake pressure is repeated until the slip value for the wheel has been brought back into the acceptable window.

This is obviously a simplified explanation of how ABS works and the algorithms are in fact very complex and will vary from one ABS implementation to another. When you remember that this algorithm must be executed on all wheels in just a few milliseconds, it is not surprising that ABS is among the most demanding microcontroller applications.

An important point worth discussing about ABS is that it is one of the most safety critical processor applications in existence. The consequences of a faulty ABS system could be potentially disastrous if the brakes were prevented from operating, or were applied erroneously. For this reason ABS manufacturers take great care in the safety aspects of the system design. It is not uncommon for two identical microcontrollers to be implemented, running the same software in parallel and continually checking each other via a communication protocol for any erroneous operation.

Another solution to this problem is to have a simpler (lower cost) slave µC, (that acts as a *watch-dog* for the main ABS microcontroller. This slave device is programmed to monitor the major activities of the master μC and it has the ability to shut down the ABS system if a fault is detected, thus reverting full braking control to the driver.

A subject worth mentioning here is traction control. Traction control is a fairly recent development and can be thought of as applying ABS in reverse. The idea of traction control is to prevent wheel-slip due to excess power on loose surfaces by applying a braking force to the slipping wheel (note that traction control is only implemented on the driven wheels). This feature is a natural progression for ABS, as all the necessary components and measurements required for traction control are inherent in the ABS system — except some means of applying a braking force when the driver is not depressing the brake pedal. This is usually achieved via an electric pump arrangement.

With the considerable improvement in safety provided by ABS, there can be little doubt that the next few years will see this system becoming more popular, possibly becoming a standard feature on all but the lowest cost cars.

The future

Hopefully this chapter will have given the reader some insight into the fascinating and challenging applications for microcontrollers in automotive applications. It has, of course, been impossible to cover all of the applications listed earlier in this chapter, or even to cover some of those we have in great technical depth (engine management or ABS themselves could each fill a text book), but the selection chosen has shown just how varied in complexity the automotive microcontroller application can be.

As a finishing thought, it may be worth pondering what the future holds for electronics, and particularly microcontrollers, in cars.

Perhaps the next major advance, one which all the major vehicle manufacturers and standards bodies are working on, is the multiplexed wiring system. As the electrical content of vehicles escalates even higher, the weight and cost of all the interconnecting cables is becoming a major concern, and the number of electrical connectors poses a reliability problem — most vehicle breakdowns are due to electrical faults. The concept of the multiplexed wiring system is to use a very high performance serial communications network between intelligent and semi-intelligent modules stationed at strategic points around the vehicle. This means that only power and the serial link need be distributed about the car — all the loads have short connections to the nearest intelligent sub-module.

The possibilities of this system are enormous; the engine management system could *talk* to the electronic gearbox controller and to the ABS/traction control system. No longer would turning on your lights simply connect power directly to the bulb — it would signal one unit to send a command to another unit, instructing it to turn on the bulb using a *Smart Power* device.

Auto electronics projects

This scenario is not fantasy, it is going to happen and because the microcontroller has a place at the heart of every one of these intelligent modules, it is safe to say that its future in the automotive market is very secure indeed.

4 Car battery monitor

Any number of things from a faulty alternator to left-on headlights (or sidelights, even) can result in a flat battery — and the first you are likely to know about it is when you turn the key one morning and the car won't start! This car battery monitor is a useful little unit designed to warn you in advance by displaying the battery's state of charge with a row of ten LEDs.

The monitor consumes a miserly 20 mA (it would take 2000 hours to discharge a 40 Ah battery), so it can be left permanently connected to the battery if required. Alternatively, it could be connected to the *ancillaries* side of the ignition switch.

Auto electronics projects

The car battery monitor will even reveal faults like a slipping fan-belt: a problem which prevents the battery from charging properly, but leaves the dashboard battery warning light off. It will even show how the battery is handling the strenuous work of starting the car (did you know it takes some twenty minutes of driving to put back what a five-second start takes out?).

Circuit

The heart of the monitor circuit (Figure 4.1) is the LM3914 bar-graph driver IC, used to drive a row of red, orange and green LEDs which together indicate a magnitude of the battery charge voltage in ten steps, approximately $1/2$ V each step from 9 V to 14 V. The IC contains an input buffer, a potential divider chain, comparators, and an accurate 1.2 V reference source. Logic is also included which gives the choice of *bar* or *dot-mode* operation — the latter is used in this application. The comparator causes the LEDs to light at 0.12 V intervals of the input voltage. TR1 acts as an amplified diode, raising the lower end of the divider chain and the negative terminal of the reference source (IC1 pins 4 and 8) to 1.9 V. The upper end of the chain at IC1 pin 6 is connected to a reference source output voltage of approximately 3.1 V from pin 7. The potential divider formed by R1 and RV1 attenuates the supply voltage and produces the signal input to the comparator, such that a supply range of 9–14 V covers the span of the divider chain and is indicated over the whole of the ten segment LED display. The LED brightness is held constant by an internal constant current source.

110

Construction

Component positions and printed circuit board track layout is shown in Figure 4.2. Construction of the project is straightforward: first fit the resistors R1 to R3 (solder and crop as you go); next insert the two presets, then fit both printed circuit board pins from the track side using a hot soldering iron to push them home. Now fit the IC socket and transistor TR1. Note that the transistor package is *not* the same as the legend on the PCB — see Figure 4.1 for pinout details. Next, identify the polarity of each LED. Hold the LED with the cathode towards you (the cathode is the shorter lead and adjacent to the flat on the lower side of the package), then with the aid of long-nosed pliers bend the leads downwards through 90 degrees at a point approximately 5 mm from the body

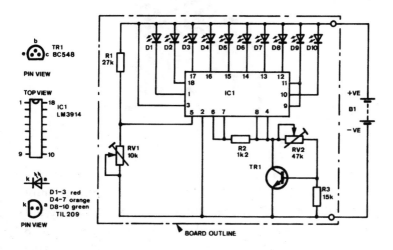

Figure 4.1   Circuit diagram

111

Figure 4.2   PCB

(see Figure 4.3). Each LED is inserted from the component side of the PCB then soldered in position to create
a line of LEDs at the same distance from the edge of the
PCB. Fit in the following order: D1–D3 red, D4–D7 orange,
D8–D10 green. Lastly insert IC1 into its socket.

112

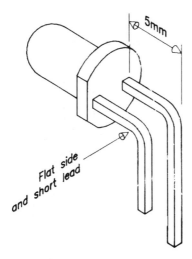

5mm

Flat side
and short lead

Figure 4.3 LED lead forming

The next job is to drill the holes in the box. Cover the
box with masking tape, as this helps with marking out
the holes and prevents scratching the box, and it also
provides a non-slip surface to prevent the drill bit moving. See Figure 4.4, for hole positions. After having drilled
all the holes, the PCB can be fitted into the box using
two M3 x 16 mm screws, with two M3 x $^1/_4$ inch spacers
under the PCB to position it at the correct height, and
the PCB secured with M3 nuts (see Figure 4.5). The zip
wire should now be soldered to the veropins; fit the P
clip to the zip wire and secure it to the M3 x 16 mm screw
using a second M3 nut. Having fitted the zip wire, insert
the fuse holder in the positive (+) supply line as close to
the battery as possible; see Figure 4.6 for assembly instructions. The fuse is included to protect the battery
wiring in the event of a short circuit. The unit is now
complete and ready for calibration.

Auto electronics projects

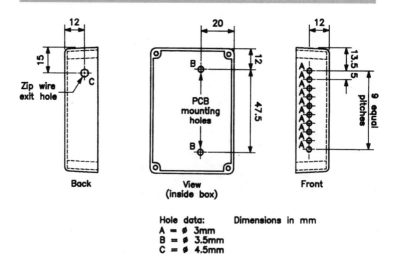

Hole data:
A = ∅ 3mm
B = ∅ 3.5mm
C = ∅ 4.5mm

Dimensions in mm

Figure 4.4   Box drilling details

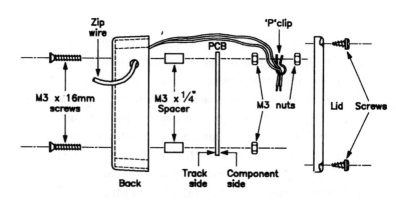

Figure 4.5   Box and PCB assembly

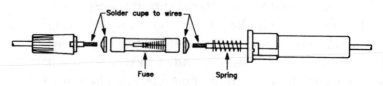

Figure 4.6   Fuse holder assembly

114

Calibration

Connect the battery monitor to a fully charged battery, or preferably use a variable voltage bench PSU set to 14 V. If this is not possible connect a battery charger to the charged battery and switch it on to ensure that a real 14 V level is achieved (maximum output from a car's charging/battery system while running is 14 V, not 12 V).

Take note — Take note — Take note — Take note

Connecting the battery monitor to the supply the wrong way round will result in permanent damage to IC1!

Set your multimeter to the 2 V range, connect the common (black) lead to 0 V, and the positive (red) lead to pin 8 of IC1. Using a screwdriver, adjust RV2 until the voltage on the multimeter reads 1.9 V. Remove the meter leads, then adjust RV1 until the top end green LED lights. The battery monitor is now calibrated. All that is left to do now is to screw the back cover onto the box, and fit it into your car. Quickstick pads are supplied with the kit to mount the box onto the dashboard if required, and remember to secure the wiring away from hot or moving parts using cable ties (order code BF91Y) as accidents can be expensive if not dangerous. Happy motoring!

Auto electronics projects

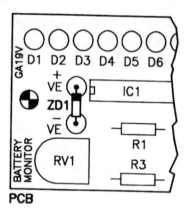

Figure 4.7   Adding zener diode protection to the module

116

Car battery monitor parts list

Resistors — 0.6 W 1% metal film

R1	27 k	1	(M27K)
R2	1k2	1	(M1K2)
R3	15 k	1	(M15K)
RV1	10 k hor encl preset	1	(UH03D)
RV2	47 k hor encl preset	1	(UH05F)

Semiconductors

D1–3	mini LED red	3	(WL32K)
D4–7	mini LED orange	4	(WL34M)
D8–10	mini LED Green	3	(WL33L)
TR1	BC548	1	(QB73Q)
IC1	LM3914	1	(WQ41U)

Miscellaneous

batt mon PCB	1	(GA19V)
DIL socket 18-pin	1	(HQ76H)
box 301	1	(LL12N)
zip wire	3 mtrs	(XR39N)
P clip $^1/_8$ in	1	(JH21X)
M3 x $^1/_4$ in spacer	1 pkt	(FG33L)
M3 x 16 mm screw	1 pkt	(JD16S)
M3 nut	1 pkt	(BF58N)
quickstick pad	1 strp	(HB22Y)

Auto electronics projects

in-line fuse holder	1	(RX51F)
1 $^1/_4$ in 100 mA fuse	1	(WR08J)
constructors' guide	1	(XH79L)
1 mm PCB pins	1 pkt	(FL24B)
instruction leaflet	1	(XK10L)
15 V 1.3 W zener diode	1	(QF57M)

All of the above available as a kit

battery monitor kit	1	(LK42V)

5   Car digital tachometer

In these days of ever-higher motoring costs the unit described here will help the driver to change gear at the most advantageous point to save fuel and extend engine life. Anyone using a car to tow a trailer or caravan will also benefit by being able to make the best use of the torque available from the engine.

Conventional tachometers give a display of engine speed on a milliammeter, usually with a scale of about 270° arc. Pulses produced by the action of the contact breakers are integrated and fed to the meter to give an analogue display of engine revolutions. The disadvantages are that (1) an average reading is displayed, which can easily lag behind rapid speed changes, and (2) meters tend to be somewhat fragile.

Auto electronics projects

The tachometer described here overcomes both of these disadvantages by counting pulses and displaying engine revolutions over a very short time, as the digital display is continuously updated. Two digits display the number of revolutions x 100, hence a display of 99 would correspond to 9,900 r.p.m. On the other hand, as the unit only has a two digit display, the reading could be in error by as much as 100 r.p.m. compared with true engine speed. The unit is designed for negative earth cars, and if you are not sure of the polarity of your car, a glance at the owners manual or even at the battery connections will tell you.

Circuit

The complete circuit is shown in Figure 5.1. Pulses produced by the make-and-break action of the engine contact breaker points are fed to IC1a, which is a dual Schmitt trigger monostable, via a resistor/capacitor network composed of R1, R2 and C1. This network helps to smooth out any high voltage spikes which may be present on the contact breaker pulses. The zener diode ZD1 limits the input pulse at IC1a to 4.7 volts, to avoid any damage to the device. To prevent any false triggering due to contact point bounce (produced when the points do not open and close cleanly) the monostable period is set to 3 milliseconds by R3 and C2, after which it is ready to be retriggered by the next pulse, and this time period allows for the maximum count for a 4-stroke, 4-cylinder engine of 10,000 r.p.m. — a speed not often attained on normal road cars! The maximum count of 10,000 r.p.m. ($\div$ 100 r.p.m.) corresponds to 20,000 pulses/minute and the time for 1 pulse is 60/20,000 seconds or 3 ms. A high

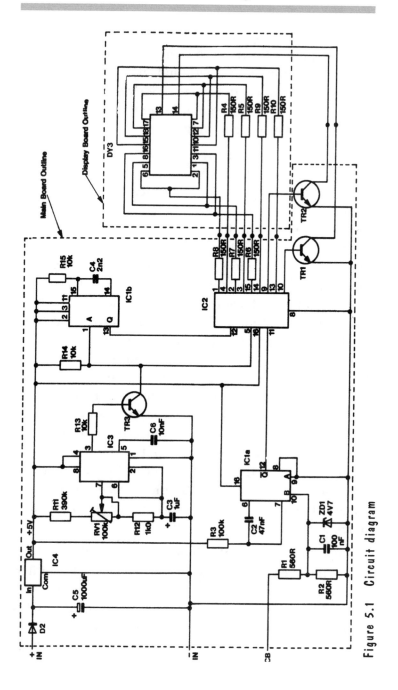

Figure 5.1 Circuit diagram

engine speed would therefore not allow enough time between pulses for triggering the monostable. This design is for 4-cylinder cars *only* and anyone using it on a 6 or 8-cylinder car would have to modify the count period accordingly, or use a compensation factor on the readings — not easy while driving!

The output pulses from IC1a, pin 12, are fed to the count input, pin 11, of IC2. This is a 4 digit counter with both latch and reset. It drives a multiplexed 2 digit display directly, with transistors TR1 and TR2 selecting the digit, and resistors R4–R10 limiting the segment current. The counter requires latch pulses in order to give sensible readings and these are provided by IC3, TR3 and their associated components. IC3 is the ever useful 555, used as an oscillator whose frequency is controlled by RV1. The oscillator output waveform, arranged such that there is a long high and a short low period, is inverted by TR3 so that a short high period is achieved. This short pulse is used to control the latch of the counter device IC2, so that when this input goes high the information in the counter is transferred to the internal latch and displayed. The trailing edge of this short pulse is used to trigger the monostable IC1b, whose output pulse is used to reset the counter so that it will start counting from *00* again. The use of a separate monostable to reset the counter ensures that the reset pulse always occurs *after* the latch pulse and a true reading displayed.

Because the supply voltage of a car, nominally 12–14 V, varies between these limits with engine speed, integrated circuit IC4 is used to derive a regulated 5 V supply from this, which is then used to supply IC1, IC2 and IC3 and is important for the stability of the oscillator (IC3). Diode D2 and capacitor C5 remove any noise on the supply.

Construction

The Digital Tachometer is constructed on two PCBs; the main board and the display board. The display board is mounted at 90° to the main board by being soldered to Veropins, and this holds the display so that it can be viewed through the cut-out display *window* at the end of the case. Figure 5.2 shows the constructional details.

First job, however, is to build up the main printed circuit board. Referring to Figure 5.3, begin by fitting the smallest components first. Check the polarity of C3, C5, and the direction of Dl, D2 before fitting, then work your way through the components by size fitting C5, the largest, last. Insert the ICs into the appropriate sockets *only*

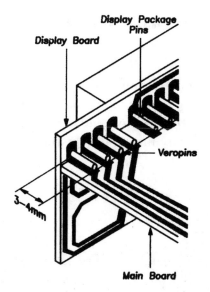

Figure 5.2   Preparing the veropins for attaching main and display PCBs together

Auto electronics projects

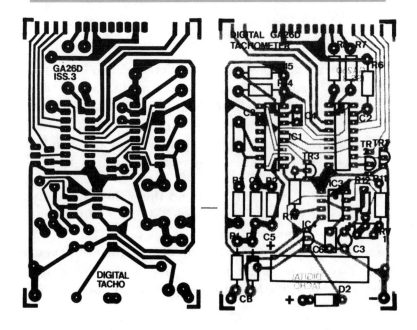

Figure 5.3 Digital tachometer main PCB and legend layout

after all other construction of the tachometer module is completed, taking the usual precautions with CMOS devices. Note that the negative end of C5 must be close to the PCB or you may find that adjusting RV1 is difficult during calibration!

Display board

Refer to Figure 5.4. First mount resistors R4, R5, R9, R10, and the veropins from the *component* side, being careful not to strip the pads off the PCB in the process! A hot soldering iron will help to push the pins home. Don't

124

forget to fit the wire link (this can be made from an off-cut from a resistor, or with single-strand wire). Solder and crop the resistors and the wire link. Next fit the display to the PCB; pins 1 to 9 are on the side where the decimal point will be found, and pins 1 and 18 are marked on the PCB. Solder and crop pins 1 to 18. Now measure the required length for the display board pins by offering the display board up to the main PCB, 3–4 mm is about right; see Figure 5.2. If the pins are too short, the connections won't be mechanically strong. After trimming the pins down you can solder the display board to the main PCB. The pins on the display board also require soldering, and if this has already been done, you may find that the two boards do not marry snugly to each other.

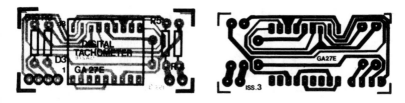

Figure 5.4   Display board layout and legend

Use mains connection wire for power supply and screened wire for the input signal soldered to the three pins. Screened cable is used to stop the emission of RF interference, and the outer screening must be connected to earth, preferably at the HT coil end. Label the function of each wire at the end that will connect to the car electrics. If you are going to use the optional box, the front panel of the case is replaced by a piece of red filter

Auto electronics projects

cut to size (using the original panel as a template) with a pair of scissors or craft knife. This slots neatly into the case, which is moulded in two sections. As you may have noticed, there is no method of mounting the tacho module into the suggested box, so the alternative is to use quickstick pads. The suggested box also needs to be modified by removing part of the battery compartment; only the top and front partitions of this need to be removed, the sides will help to keep the PCB central in the box, and the screw holes must remain intact or else the box cannot be fastened together (see Figure 5.5).

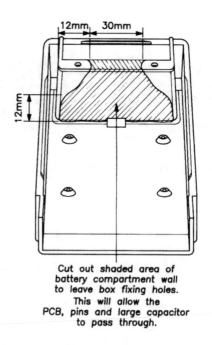

Cut out shaded area of
battery compartment wall
to leave box fixing holes.
This will allow the
PCB, pins and large capacitor
to pass through.

Figure 5.5   Box modification details

Setting up

One advantage of a digital tachometer over an analogue type is the ease of setting-up and calibration. Only one adjustment to RV1 needs to be made and, barring accidents, will prevail for the life of the unit. This setting ensures that the oscillator runs at the correct frequency and the method of calibration depends on the equipment available. Calibration against another tachometer is possible, setting RV1 to give a display of 30 when the standard tachometer reads 3000 r.p.m. If you have access to a signal generator, set the frequency to 100 Hz (sine or squarewave) and the output level to maximum (more than 4.7 V). Connect this signal to the I/P pin on the PCB, and again this will simulate an ignition pulse train input of 3000 r.p.m.

Alternatively calibration can be carried out against the mains frequency by using a transformer and bridge rectifier to provide a 100 Hz signal as shown in Figure 5.6,

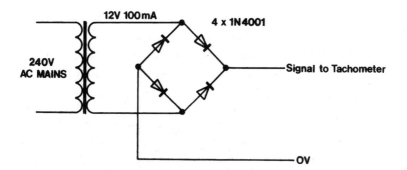

Figure 5.6   Mains frequency doubler for calibration

Auto electronics projects

and a battery charger is very effective in this role. In either case RV1 is adjusted to give a display reading of 30. Calibration should include a test run for up to half an hour or so for *warming up* and stabilisation, whereafter it might be noted that RV1 requires further fine tuning. When you are satisfied with the calibration of the counter, RV1 should be fixed in position with wax, paint or nail varnish.

Fitting the unit into the car

After calibration, the unit is ready to be fitted to the car. It is impossible to give detailed instructions for every car but the following notes may be helpful.

● it is a good idea to try the unit in various positions for best readability, using adhesive tape, until you are satisfied,

● having decided on the best position use double-sided tape, adhesive pads or two pieces of velcro-tape, one glued to the unit and one glued to the car dashboard. All of these methods, of course, mean that the unit can be removed easily and the dashboard cleaned and left unmarked,

● alternatively, use self-tapping screws through one half of the case into the dashboard. This works well, but unless you can utilise existing screw holes you will be left with holes in the dashboard if you decide to remove the unit.

Car digital tachometer

The three leads must pass into the engine compartment and it is important that they are protected by a rubber or plastic grommet. It may be possible to squeeze them through an existing cable entry or you may have to drill a new hole, but either way make sure they are protected. Connection to the car electrics is fairly straightforward; the tacho input lead is connected to the CB terminal of the HT coil, which can be identified by the lead from the points and distributor to the HT coil. *CB* stands for *contact breaker*, often marked with a (–) minus sign. The supply would best be taken from the ignition switch via a 100 mA fuse, so that the unit is switched off when the car is not running. The easiest way of doing this would be to follow the other lead from the HT coil (marked with a (+) plus sign) up to the ballast resistor (if fitted), and make the connection to the other side of it, see Figure 5.7.

Take note — Take note — Take note — Take note

Not all ignition systems are the same so consult your workshop manual before attempting to fit the tacho. Also please remember that a car engine compartment is a hazardous area — never attempt to fit the tacho, or anything else for that matter, while the engine is running! Also, secure all cables away from hot or moving parts! Anchor them to existing wiring looms using cable ties.

Auto electronics projects

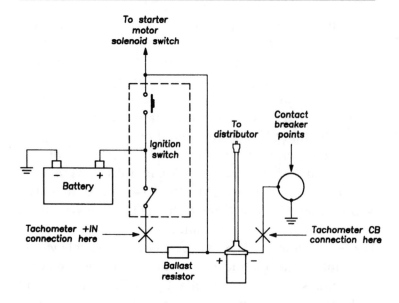

Figure 5.7 Connecting tachometer to a typical ignition system with contact breaker

Car digital tachometer parts list

Resistors — All 0.6 W 1% metal film

R1,2,	560 Ω	2	(M560R)
R3	100 k	1	(M100K)
R4–10	150 Ω	7	(M150R)
R11	390 k	1	(M390K)
R12	1 k	1	(M1K)
R13–15	10 k	3	(M10K)
RV1	100 k vert encl preset	1	(UH19V)

Capacitors

C1	100 nF polyester	1	(BX76H)
C2	47 nF poly layer	1	(WW37S)
C3	1 µF 35 V tantalum	1	(WW60Q)
C4	2n2F mylar	1	(WW16S)
C5	1000 µF 16 V axial	1	(FB82D)
C6	10 nF 50 V disc	1	(BX00A)

Semiconductors

IC1	74LS221	1	(YF86T)
IC2	74C925	1	(QY08J)
IC3	NE555	1	(QH66W)
IC4	LM78L05ACZ	1	(QL26D)
TRl–3	BC549	3	(QQ15R)
ZD1	BZY88C4V7	1	(QH06G)
D2	1N4001	1	(QL73Q)
DY3	DD display type C	1	(BY68Y)

Auto electronics projects

Miscellaneous

8-pin DIL socket	1	(BL17T)
16-pin DIL socket	2	(BL19V)
dig tacho main PCB	1	(GA26D)
dig tacho display PCB	1	(GA27E)
1 mm PCB pin	1 pkt	(FL24B)
red display filter	1	(FR34M)
cable grommet	1	(LR47B)
twin mains DS black	3 mtrs	(XR47B)
cable single black	3 mtrs	(XR12N)
quickstick pads	1	(HB22Y)
constuctors' guide	1	(XH79L)
instruction leaflet	1	(XK03D)

All of the above available as a kit

car digital tachometer kit	1	(LK79L)

Optional (not in kit)

small remote control box	1	(LH90X)
in-line fuse holder	1	(RX51F)
$1\,{}^{1}/_{4}$ in 100 mA fuse	1	(WR08J)
velcro mounts	as reqd	(FE45Y)
cable tie-wrap 100	as reqd	(BF91Y)

6 Car lights-on warning indicator

If your car is not fitted with some kind of lights-on warning, the chances are that you will at some time (if you have not already done so!) leave your lights switched on. Murphy's law dictates that when you do so, your absence from the car will be of sufficient duration to ensure that the battery will be well and truly flat. Of course Murphy, not content to do things by halves, will ensure that it happens when you are late for some important occasion and that there is no one else around to give you a push or a jump start!

Modern cars further aggravate the situation as many of them, being fitted with electronic ignition or electronic engine management systems, just plain refuse to be push-started!

Auto electronics projects

It is amazing that such mechanically advanced cars often *do not* have a lights-on warning indicator of some kind. To illustrate this, the prototype was installed in a 2.0 litre petrol-injection Ford Sierra Estate — despite being a *Ghia*, there was no lights-on warning device!

Various warning devices are available, however, some become a nuisance because they sound continuously when the lights are deliberately left on. For instance, while the driver is waiting in the car at night, with the engine switched off.

Some more sophisticated devices will not sound if the lights are switched on again *after* the ignition has been switched off, i.e. for parking lights. However, this fails to warn the driver if he inadvertently *knocks* the light switch on when leaving the car — as could be the case with many cars having the light switch *stalk* on the driver's door side of the steering column.

This lights-on warning indicator will emit a clearly audible buzzing sound when the car lights are left on, the ignition switch is turned-off and the driver's door opened. In this manner the buzzer will only sound when the driver is genuinely about to leave the car.

Now that you are thoroughly convinced that for the sake of a few pounds, you need not be caught out in the future, why not build this handy accessory (which the manufacturer should have included as standard) and fit it into your car? Enterprising readers may wish to offer this *add-on* to friends, relatives and neighbours for a suitable fee (don't forget to tell the tax man!). A personal tale of woe and the assurance that, "*I've got one and it has stopped me from getting caught out again!*" is sure to win a few favourable responses.

134

Car lights-on warning indicator

Circuit description and operation

The circuit of the lights-on warning indicator is very simple, as can be seen from Figure 6.1. However, it is worthwhile to know how the circuit operates as this will help, should problems occur.

P1 of the unit is connected to the sidelight circuit of the car and provides power to the circuit only when the lights are switched on. The sidelight circuit is live when either sidelights or headlights are switched on.

P2 is connected to the accessory circuit and when the ignition switch is off, P2 is pulled low via resistor R3 (P3 is connected to 0 V). Diode D1 is forward biased and turns on transistor TR1 via resistor R2. Note that the internal resistance of accessories (i.e. radio-cassette) may be sufficiently low to make the connection to P3 unnecessary; this can be determined by experimentation.

P6 is connected to the driver's door switch, thus when the door is opened, a complete path to 0 V is provided by the door switch, allowing the buzzer to sound.

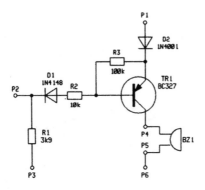

Figure 6.1   Circuit diagram

135

Auto electronics projects

When the ignition switch is on, P2 is pulled high, reverse biasing diode D1. Resistor R3 ensures that transistor TR1 is held in the off state. The positive supply to buzzer BZ1 is removed and thus prevents it from sounding, regardless of whether the driver's door is open or shut.

When the lights are off and the car doors are closed, the polarity of the supply to the unit is effectively reversed. Diode D2 prevents damage to the circuit under this condition.

Construction

Assembly of the unit is simplicity itself. Referring to Figure 6.2, it is advised that the PCB pins are fitted first, followed by the resistors and the diodes and finally the transistor. Make sure that the transistor is fitted fairly close to the PCB otherwise the PCB will not fit into the case.

Next solder the buzzer's wires to the PCB pins, red (+V) to P4, black (–V) to P5. Attach the connecting wires to the PCB pins and label the free ends so that you can identify the wires after the PCB has been fitted into the case!

The PCB simply lies in the case, the wires protruding through the aperture provided. Screw the case together

Figure 6.2   PCB legend and track

and affix the buzzer onto the lid of the case using one of the double-sided adhesive pads. The other pads can be placed onto the underside of the case ready for fitting into the car.

Although it is unlikely that there will be any problems with the unit, it is advisable to test it before fitting into the car. It is easier to take remedial action on the work bench than underneath the car dashboard! Using a 9 to 14 V supply (i.e. PP3-sized battery, battery eliminator, etc.) connect P3 and P6 to 0 V, then connect P1 to +V, the buzzer should sound. Connect P2 to +V as well, this should silence the buzzer.

Installation

Refer to Figures 6.3, 6.4, 6.5 and 6.6. It is necessary to gain access to the car's wiring, which will undoubtedly involve removing the underside of the dashboard, trim

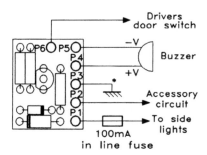

* Chassis connection
  may not be required
  See text

Figure 6.3 PCB connections

Auto electronics projects

panels, etc. It is advisable to refer to a workshop manual, e.g. of the *Haynes* variety; if you do not have one, either buy one — as it will be useful anyway, or borrow one from your local library. A workshop manual will also help you to ascertain the correct wires to connect to — otherwise it will be a case of tracing the correct wires with a multimeter.

Take note — Take note — Take note — Take note

Disconnect the car battery before making connections to the wiring. Connections to existing wiring can be made using *snap lock* connectors or terminal blocks of adequate current rating — remember the lights-on unit draws very little current, but two 55 W headlamp bulbs draw considerably more! Ensure that the new wiring will not become entangled with any controls, especially the brake pedal and steering column. To prevent short circuits, make sure that all connections are properly insulated, use adhesive electrical tape.

Connect P1, via a fuse and fuseholder, to a point in the wiring which becomes live when the sidelights are switched on (Figure 6.4).

Connect P6, to the driver's door switch (Figure 6.5). To prevent other doors from operating the buzzer, install an MR751 diode in series with the wire to the courtesy light.

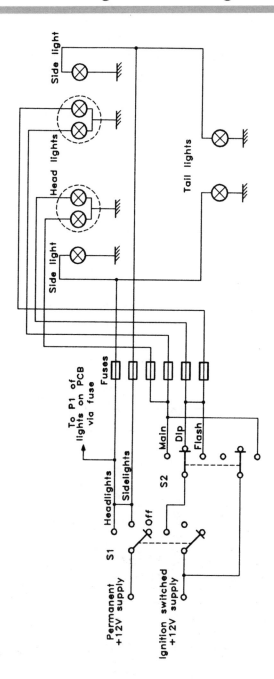

Figure 6.4 Typical lighting circuit and connections

Auto electronics projects

Connect P2 to a point in the wiring which becomes live when the ignition switch is turned to *accessory*, i.e. +V supply to the radio (Figure 6.6). Alternatively, if there is no *accessory* position, connect P2 to a point in the wiring which becomes live when the ignition switch is turned to *ignition*.

Connect P3 to the car's chassis (0 V) or to a point in the wiring which is permanently connected to the car's chassis. Note that this connection may be unnecessary if the internal resistance of any accessory is sufficiently low. This may be ascertained by testing the unit with P3 left unconnected and all accessories switched *off*. If in doubt connect P3 as previously described.

Double-check connections, reconnect the car battery.

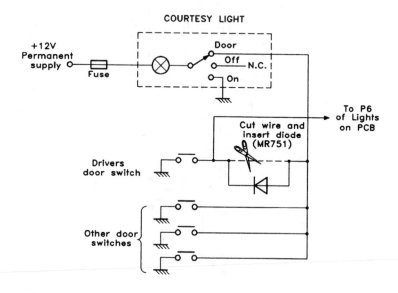

Figure 6.5   Typical ignition switch circuit and connections

Car lights-on warning indicator

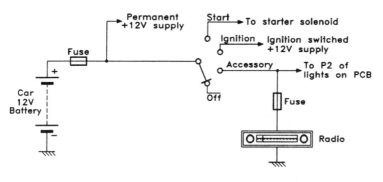

Figure 6.6 Typical ignition switch circuit and connections

Testing

●   switch lights on, leave ignition switched off and open the driver's door; the buzzer *should* sound,

●   with the driver's door shut, opening any other door should *not* cause the buzzer to sound,

●   with the ignition switched to *accessory* or *ignition*, opening the driver's door should *not* cause the buzzer to sound,

●   with lights turned off, the buzzer should *not* sound with any combination of ignition switch positions or doors open or closed.

Assuming the unit is working correctly, refit underside of dashboard and trim panels. Happy motoring.

141

Auto electronics projects

Lights-on warning parts list

Resistors — all 0.6 W 1% metal film

R1	3k9	1	(M3K9)
R2	10 k	1	(M10K)
R3	100 k	1	(M100K)

Semiconductors

D1	1N4148	1	(QL80B)
D2	1N4001	1	(QL73Q)
TR1	BC327	1	(QB66W)
	MR751	1	(YH96E)

Miscellaneous

BZ1	12 V buzzer	1	(FL40T)
P1–6	1 mm PCB pins	1 pkt	(FL24B)
	1 1/4 in in-line fuseholder	1	(RX51F)
	1 1/4 in 100 mA fuse	1	(WR08J)
	PCB	1	(GE88V)
	mini box and base	1	(JX56L)
	quickstick pads	1 strp	(HB22Y)
	instruction leaflet	1	(XT11M)
	constructors' guide	1	(XH79L)

All the above available as a kit

	lights-on warning kit	1	(LP77J)

Optional (not in kit)

16/0.2 wire	as req	(FA26D-FA36P)
snap lock cable connector	as req	(JR88V)
5 A terminal block	as req	(HF01B)

142

7 Courtesy lights extender

How many times have you got into your car of a night only to find the ignition switch has gone for a walk around the dashboard, as your aimless efforts to start the car only result in the ignition key gouging several grooves into the plastic?

This project keeps the interior light on after the car door has been closed, allowing time to find keys, ignition switch, or even your way out of the garage!

Circuit description

Figure 7.1 shows the circuit diagram of the courtesy light extender. P1 and P3 connect directly across a door switch

Auto electronics projects

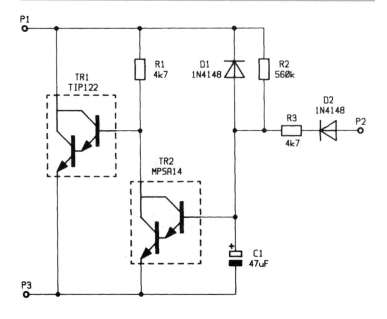

Figure 7.1 **Circuit diagram**

controlling the interior light and P2, to a source that has power while the ignition is on.

With a door open, P1 and P3 are effectively shorted, causing capacitor C1 to discharge through diode D1. As soon as all doors are closed, as capacitor C1 is discharged, transistor TR2 is turned off. Resistor R1 pulls the base of TR1 high, turning it on and causing current to flow through the courtesy light. Capacitor C1 now starts to charge through resistor R2 until transistor TR2 turns on, pulling the base of transistor TR1 low and halting the flow of current through the interior light. Because capacitor C1 charges through resistor R2 and the courtesy light, it can be seen that the wattage of the interior light plays quite an important role in the time delay given.

144

Figure 7.2 shows typical time delays given at various values of R2 for 5, 10, 15 and 20 watt courtesy lights.

If, during the time delay given by the unit, the ignition is turned on, capacitor C1 charges up *very* quickly through resistor R3, turning the interior light off almost immediately. This avoids the possibility of driving away at night with the courtesy light still on.

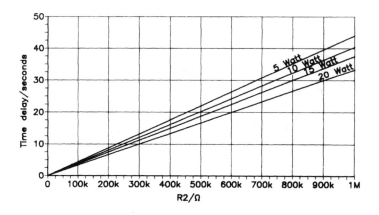

Figure 7.2 Graph of R2 against time delay for various wattage courtesy lamps

Construction

The postage stamp sized printed circuit board is of the high quality, single-sided glass fibre type, see Figure 7.3. The sequence in which the components are fitted is not critical; however, the following instructions will be of use in making these tasks as straightforward as possible.

145

Auto electronics projects

Insert and solder the PCB pins using a hot soldering iron. If the pins are heated, very little pressure is required to press them into position. Once in place, the pins may then be soldered. It is now easier to start with the smaller components, such as the resistors, work upwards in size, and transistor TR1 is fitted last.

The diodes should be inserted such that the band at one end of the diode corresponds with the white block on the PCB legend. When fitting the electrolytic capacitor, it is essential that the correct polarity is observed. The negative lead of the capacitor, which is usually marked by a full-length stripe and a negative (–) symbol, should be inserted away from the hole marked with a positive (+) sign on the PCB legend. Insert and solder the two transistors, matching the shape of each case to its outline on the legend.

Lastly, solder lengths of wire to the veropins and mount the PCB inside the box, as shown in Figure 7.4.

Installation

The courtesy light extender is extremely simple to fit. However, for someone who is not familiar with automotive electrical installation it is advised that they seek the advice of a qualified person before proceeding.

There are two methods of switching the interior courtesy light:

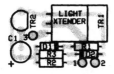 

Figure 7.3   PCB legend and track

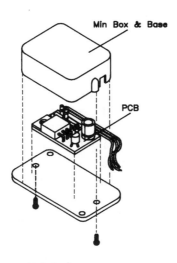

Figure 7.4   Fitting unit into box

Take note — Take note — Take note — Take note

When carrying out any form of electrical work on a vehicle always disconnect the battery and never work inside the engine compartment with the engine running!

Auto electronics projects

●   door switches are fitted to the 0 V side of the courtesy light, for installation follow Figures 7.5(a) or 7.5(b),

●   door switches are fitted breaking the +12 V supply to the courtesy light, for installation follow Figures 7.5(c) or 7.5(d).

In its simplest configuration, the unit connects directly across a door switch; P1 connecting to the more positive side of the switch and P3 to the more negative.

If ignition override is required then P2 must be connected to a source that has power while the ignition is on (for example, + SW terminal of the ignition coil). If no easy connection can be made to the ignition circuit then P2 can be connected into the *accessory* circuit.

As the complete unit is small and unobtrusive it can easily be mounted inside a door-post, behind an existing door switch. The box can be held in place using a self-adhesive pad (such as HB22Y) or bolted down using the two mounting holes provided in the base of the box. Check behind panels before drilling any holes and ensure that no wiring harness or other components are located behind panels that would otherwise be damaged.

	Min	Typ	Max	Units
Operating voltage	10	12	15	V
Quiescent current at 12 V		3		mA
Maximum switching current			5	A

Table 7.1   Specification of prototype

Courtesy lights extender

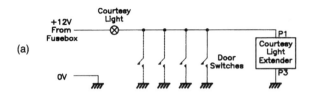

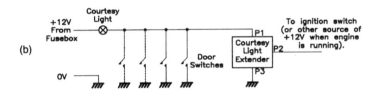

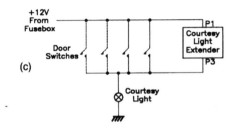

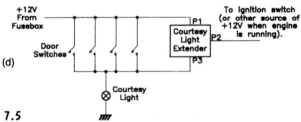

Figure 7.5

(a) Simple connection for vehicles with negative switched courtesy light

(b) Connection for negative switched courtesy light with ignition override

(c) Simple connection for vehicles with positive switched courtesy light

(d) Connection for positive switched courtesy light with ignition override

149

Auto electronics projects

Courtesy light parts list

Resistors — All 0.6 W 1% metal film

R1,3	4k7	2	(M4K7)
R2	560 k	1	(M560K)

Capacitors

C1	47 µF 16 V minelect	1	(YY37S)

Semiconductors

TR1	TIP122	1	(WQ73Q)
TR2	MPSA14	1	(QH60Q)
D1,2	1N4148	2	(QL80B)

Miscellaneous

P1,2,3	1 mm PCB pins	1 pkt	(FL24B)
	PC board	1	(GE81C)
	min box and base	1	(JX56L)
	courtesy light leaflet	1	(XK96E)
	constructors' guide	1	(XH79L)

All of the above are available as a kit

courtesy light kit	1	(LP66W)

Optional (not in kit)

16/0.2 black hook up wire	1	(FA26D)
16/0.2 red hook up wire	1	(FA33L)

8 Car audio switched-mode psu

Auto electronics projects

For many years the motorist has not been able to benefit from hi-fi quality sound while travelling in the car. For the long-distance traveller, business executive or hi-fi buff on-the-move, the car is a far from ideal environment for listening to music; this is due to a number of reasons. First, the car's interior is designed for conveying passengers and not for ideal location of conventional *box design* loudspeakers. Second, the sound replay/receiving equipment has to be miniaturised and capable of operation in a very harsh environment. Dashboard temperatures often exceed 60°C in hot weather (yes, even in the English climate!) and fall to several degrees below zero in cold weather. Vibration and humidity also add to the stresses that the equipment must endure. Third, the low, noisy and somewhat variable supply voltage makes life even more difficult for the electronic circuitry.

The environmental and size problems of the car environment have largely been solved by cleverly designed equipment. Car loudspeakers are optimised for operation in rear parcel shelves and door panels instead of conventional sealed or ported enclosures. Car radio, cassette, CD (compact disc) and DAT (digital audio tape) equipment is very compact. Such equipment is designed for either mounting in the dashboard/centre console or remote mounting in the boot or under a seat, with just the controls located within the driver's easy reach.

It is however, the third point that is the main reason for this project, the vehicle electrical supply. The 12 V electrical system is far from ideal when it comes to powering audio amplifiers. The electrical system itself, although generally referred to as being 12 V, usually operates at around 13–14 V when the engine is running. By convention, the voltage when the engine is running is assumed to be 13.8 V.

A singled-ended amplifier operating from a supply voltage of this (low) level is capable of delivering around 7 W r.m.s. into a 4 Ω load. If a BTL (bridge tied load) amplifier is used the power output can be increased to around 22 W r.m.s. into a 4 Ω load. Most *high power* radio/cassette players have an output power of around 22 W r.m.s., regardless of how many watts the advertising brochures boast!

For hi-fi quality sound reproduction in a car it is necessary to have the capability of higher power levels. This not being required for *blowing out the windows* (although often used as such by drivers of ageing Ford Cortinas with pink fluffy dice hanging from the interior rear-view mirror), but simply because a high power amplifier operating at modest power levels will introduce far less distortion and handle transients far better than a medium power amplifier running almost flat out. This is especially true if the sound source is CD, where the dynamic range of the recording is often very wide.

There are two ways in which the output power can be increased, by either decreasing the loudspeaker impedance or increasing the supply voltage. The main disadvantage of the former method is that car speakers are not commonly produced with impedances below 4 Ω and that power losses in cables are increased. The latter method of increasing the supply voltage is commonly used in high power car *boosters* and in hi-fi car audio amplifiers — this is the method that is described here.

Auto electronics projects

Circuit description

Figure 8.1 shows a block diagram representation of the power supply and Figure 8.2 shows the full circuit diagram.

The supply input to the power supply is via P1 (+V) and P2 (0 V). The power supply is connected directly to the vehicle battery via high current cables, therefore the off-board supply fuse FS1 is essential in case of a fault causing a short circuit directly across the battery. Remote power switching is achieved by TR1, RL1 and associated components. The control input P3, when taken to +V, biases TR1 on and operates RL1, thus powering-up the rest of the supply. LD1 serves to indicate *power on*. The control signal is provided by the *electric aerial* output found on most radio-cassette units. Diode D1 clamps the voltage spike produced by RL1 when it de-energises. Diode D2 provides polarity protection by blowing fuse FS2 and preventing the remote power switch from operating.

Capacitors C1, C2, C3, C4, C5 and toroid L1 form the input π-filter, the output of which supplies the push-pull output stage. The power MOSFETs are arranged in two pairs (TR2 and TR3) and (TR4 and TR5), each driving one half of the transformer primary. Resistor R8 and capacitor C6 form a snubber network to increase the rise-time of switching spikes. Components ZD1, D3 and TR2 and ZD2, D4 and TR5 form an active spike clamp, employed to protect the MOSFETs' drain/source junctions from high voltage switching spikes. This operates by feeding the spike back into the gate of the relevant

154

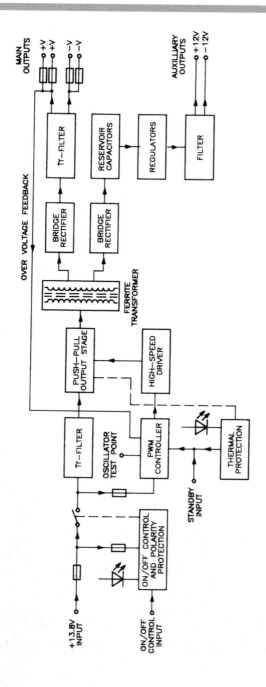

Figure 8.1 Block diagram of the switching PSU

Auto electronics projects

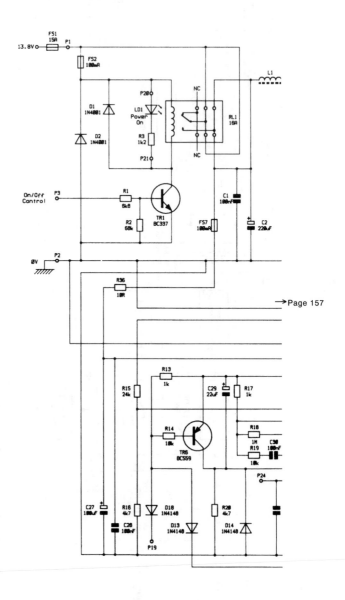

→Page 157

Figure 8.2   Circuit diagram of the switching PSU

Car audio switched-mode psu

→Page158

←Page 156

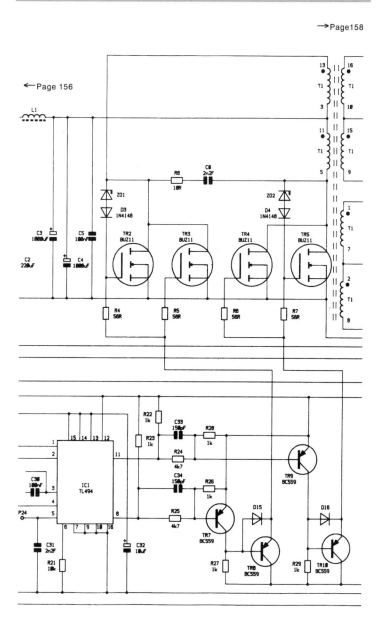

Figure 8.2 Continued

Auto electronics projects

←Page 157

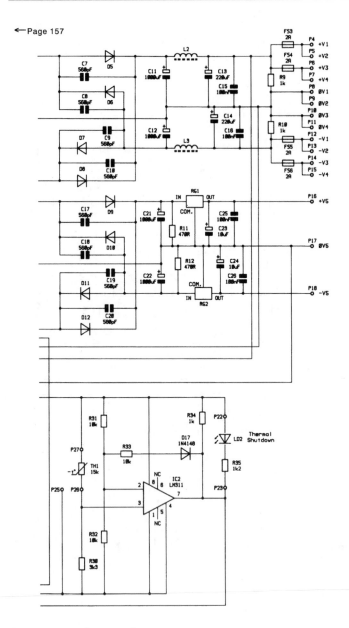

Figure 8.2   Continued

MOSFET thus turning it on and clamping the spike. Gate resistors R4 to R7 help to balance current flow through each MOSFET pair and also help to reduce switching noise.

T1 is a step up transformer comprising six windings, two connected to form a centre tapped primary winding and four are connected in two pairs to form two centre tapped secondary windings.

Components R36, C27 and C28 form a simple R–C filter for IC1 which attenuates supply borne noise. C29 and R20 set the soft-start time period for IC1. At switch on, C29 is discharged and IC1's outputs are inhibited. As C29 charges via R20, the pulse width of the PWM drive signals are allowed to increase from zero. D14 prevents IC1's soft-start input from being pulled negative at switch-off and also serves to discharge C29 more quickly. TR6 discharges C29 and inhibits IC1's outputs in response to a thermal shutdown condition or a standby input (low) from P19. D18 and D13 form a discrete AND circuit. When the shutdown condition and standby inputs are removed, TR6 allows C29 to charge again and the power supply restarts.

Resistor R21 and capacitor C31 set the oscillator frequency, P24 may be used to monitor the oscillator waveform. Care should be exercised to ensure that this pin is not subject to undue capacitive loading, otherwise the oscillator frequency will shift.

Resistors R17, R18, R19 together with capacitor C30 form a phase selective network that sets the gain of the over-voltage amplifier. Phase compensation is necessary to ensure good loop stability, otherwise the power supply

Auto electronics projects

could break into oscillation. Resistors R15 and R16 form a potential divider which is used to apply over voltage feedback to IC1, with the values as shown, the maximum output voltage is ±30 V.

Transistors TR7 to TR10 and associated components form two high speed driver circuits which are able to charge and discharge the gate capacitance of each of the MOSFETs very quickly. Circuit operation for one of the (two identical) drivers is as follows: R23 is the pull-up resistor for the open collector output of IC1 (pin 8). When pin 8 goes low (output on) TR7 is biased on by R25 (C34 serves to increase switching speed), D15 conducts and TR2, TR3 turn on quickly. At this time TR8 is switched off. When IC1 pin 8 goes high (off) TR7 switches off and TR8 base is pulled low; as the gates of TR2 and TR3 are charged to a positive potential, D15 is reverse biased and TR8 conducts. This action rapidly switches off TR2 and TR3.

Integrated circuit IC2 is a comparator with its inputs connected to two potential dividers. Resistors R31 and R32 form a reference potential divider and thermistor TH1 and R30 form a temperature sensing network. R33 and D17 provide a large degree of hysteresis when the output changes state. Normally the output from IC2 (pin 7) is high and the voltage on pin 2 is around $1/2$ supply. The voltage on pin 3 is dependent on the resistance of TH1, governed by the heatsink temperature with which it is in contact. As the temperature of the heatsink rises, the resistance of TH1 reduces and the voltage on pin 3 increase. When the voltage on pin 3 exceeds the voltage on pin 2, the output of IC2 goes low. LD2 illuminates indicating thermal shutdown and the power supply shuts down. At this point D17 conducts, this adds R33 to the lower half of the reference divider reducing the reference potential on pin 2 to around $\frac{1}{3}$ supply (ignoring D17 voltage drop and saturated output voltage of IC2). The voltage on pin 3 will now have to fall below $\frac{1}{3}$ supply before the circuit will reset and the supply allowed to restart. Correspondingly the resistance of TH1 will have to rise and its temperature fall before supply operation is resumed. With the circuit values as shown, the trip temperature is 80°C and the reset temperature is 60°C.

Diodes D5 to D8 form a bridge rectifier (main output), the devices used are high speed types, essential for use in switch mode applications. Capacitors C7 to C10 help to reduce transients and switching noise. Components C11, C12, L2, L3, C13, C14, C15 and C16 form π-filter networks for the main outputs. Resistors R9 and R10 serve to provide a *minimum load* for the power supply and also discharge the filter capacitors quickly after switch-off. Fuses FS3 to FS6 provide protection against short circuits and overloads. Positive 30 V outputs are available from P4, 5, 6 and 7. Negative 30 V outputs are available from P12, 13, 14 and 15. Pins 8, 9, 10 and 11 provide a zero volt return.

Diodes D9 to D12 form a second bridge rectifier (auxiliary output), again high speed types are used. Capacitors C17 to C20 help to reduce transients and switching noise. Capacitors C21 and C22 are the reservoir capacitors for the auxiliary output. Resistors R11 and R12 serve the same purpose as R9 and R10 in the main output circuitry. Voltage regulators RG1 and RG2 regulate the supply rails and attenuate switching noise on the auxiliary output.

Auto electronics projects

Capacitors C23, C24, C25 and C26 are decoupling capacitors and ensure supply stability. Positive and negative 12 V auxiliary outputs are available on P16 and P18 respectively. P17 provides a 0 V return.

Construction

The PCB is of the single-sided glass fibre type, with a printed legend to assist insertion of the components. To increase the current rating of some of the tracks it is necessary to tin the exposed areas of track on the underside of the PCB. These tracks will be clearly seen as they are not covered by the solder resist layer. Tinning of the tracks should actually be the final assembly task. Removal of misplaced components can be very difficult, especially on a densely populated board such as this, so please double check component type, value and orientation (where appropriate) before inserting and soldering the component.

Referring to the following constructional notes, the parts list and Figure 8.3, begin construction. It is recommended that the following construction order is adhered to closely, otherwise it will be found extremely difficult, to fit some of the components.

Start by inserting the three 22 SWG wire links, these are indicated on the PCB by a single straight line and an adjacent *LK* mark.

Next insert the 1N4148 signal diodes, ensuring correct orientation.

Insert 0.6 W metal film resistors, but do not insert the 3 W wire wound resistors at this stage.

Bend and insert the four 16 SWG wire links, these are indicated on the PCB by a single straight line and an adjacent *LK number*.

Next insert the 1N4001 diodes and the two 39 V zener diodes.

Referring to Figure 8.4, loosely fit the M3 power input connection hardware and solder the M3 nuts to the PCB pads.

Insert the polystyrene capacitors and the ceramic capacitors.

Next insert the DIL sockets, but do not insert the ICs at this stage.

Insert the 45 PCB pins into the holes for TR2 to TR5 and D5 to D8; and positions marked with a circle and a *P number*. Do not insert pins into positions marked with a circle and a *W number*.

Next insert the fuse clips. You should find that by carefully bending over the two legs on the track side of the PCB before soldering, the fuse clips will remain straight.

Insert the BC337 and BC559 transistors, ensuring correct orientation.

Next insert the tantalum capacitors, ensuring that the correct voltage rating capacitor is inserted in the correct location. Tantalum capacitors are polarised and must be correctly orientated, the plus (+) sign on the body must be inserted into the hole nearest that marked with a plus sign.

Auto electronics projects

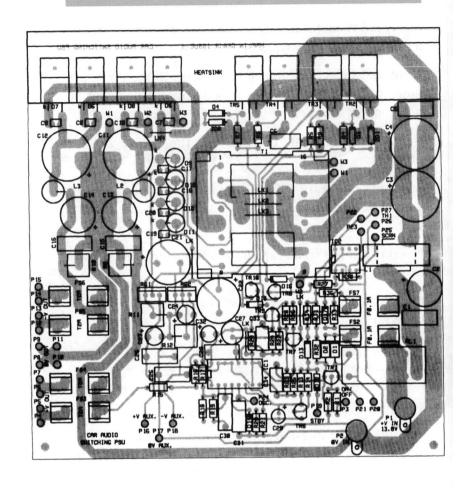

Figure 8.3   PCB track and legend

164

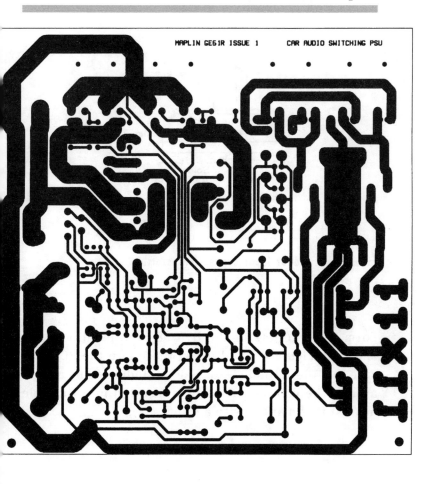

Figure 8.3   Continued

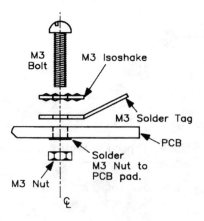

Figure 8.4   Power input connection assembly

Form the leadouts of the BYW98 rectifier diodes, as shown in Figure 8.5 and insert these into the PCB. Ensure that the cathode lead, which is indicated by a band around the component body is inserted into the hole nearest that marked with a k sign.

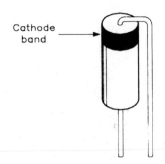

Figure 8.5   Lead formation for BYW98 rectifiers

166

Insert the 0.1 µF poly layer capacitors and the small electrolytic capacitors. The electrolytic capacitors are polarised and must be correctly orientated, the negative (–) stripe on the capacitor can must be inserted into the hole furthest away from the hole marked with a plus (+) sign.

Drill the heatsink as shown in Figure 8.6. Form the leads of the BUZ11 MOSFETs and the BYW80 rectifiers as shown in Figures 8.7 and 8.8. Assemble the heatsink assembly

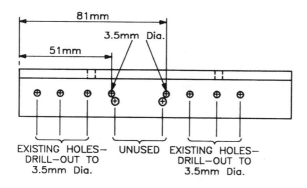

Figure 8.6   Heatsink drilling information

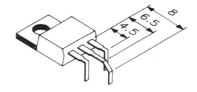

Figure 8.7   Lead formation for BUZ11 MOSFETs

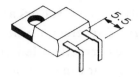

Figure 8.8   Lead formation for BYW80 rectifiers

using the M2.5 hardware as shown in Figure 8.9 and Photo 8.1. Solder the leadouts of the transistors and rectifiers to the PCB pins. Referring to Figures 8.10 and 8.11 and Photo 8.1, connect the screened cable to the thermistor and the PCB pins, use heat shrink sleeving where necessary to avoid short circuits. Glue the thermistor to the heatsink using some epoxy resin. Hold the thermistor in place whilst the resin sets.

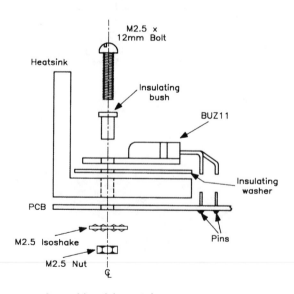

Figure 8.9   Assembly of heatsink components

Car audio switched-mode psu

Photo 8.1   Close-up of heatsink assembly

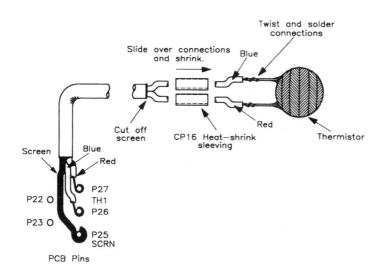

Figure 8.10   Thermistor connection

Auto electronics projects

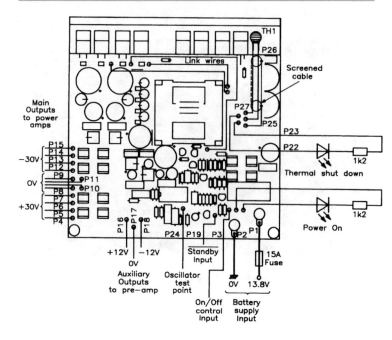

Figure 8.11 Wiring to switching PSU

Insert the two regulators, ensuring that the correct type is fitted in the correct location and that the package lines up with the outline marked on the legend. Ensure that the two metal tabs do not touch, see Photo 8.2.

Referring to Figures 8.12 and 8.13 extend the leadouts of the 3 W resistors and axial inductors and insert these into the PCB, see Photo 8.2.

Referring to Figure 8.14 wind $2^1/_2$ turns of two lengths of 16 SWG EC wire wound bifilar (side by side) around the toroid core. Prepare the ends of the EC wire to facilitate soldering and insert this inductor into the PCB at the

Car audio switched-mode psu

Photo 8.2   Close-up of the regulators, resistors and inductors

position marked L1, see Photo 8.3. It is helpful to smear the windings and toroid core with silicon rubber sealant to prevent the assembly from rattling.

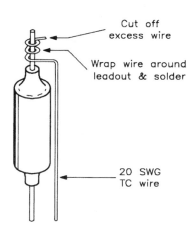

Cut off
excess wire

Wrap wire around
leadout & solder

20 SWG
TC wire

Figure 8.12   Extending leadouts of axial inductors

Auto electronics projects

Cut off
excess wire

Wrap wire around
leadout & solder

20 SWG
TC wire

Figure 8.13   Extending leadouts of axial resistors

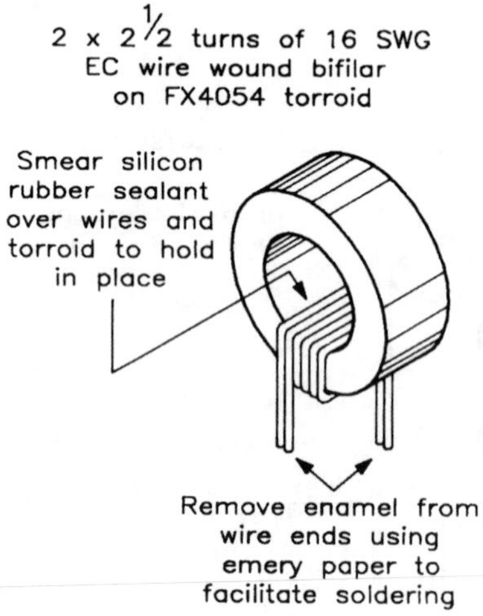

$2 \times 2\frac{1}{2}$ turns of 16 SWG
EC wire wound bifilar
on FX4054 torroid

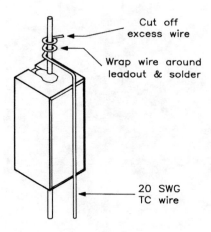

Smear silicon
rubber sealant
over wires and
torroid to hold
in place

Remove enamel from
wire ends using
emery paper to
facilitate soldering

Figure 8.14   L1 winding information

Photo 8.3   Close-up of the toroid inductor fitted into the PCB

Next insert the power relay.

Insert the large SMPS electrolytic capacitors, ensuring that the correct voltage rating capacitors are fitted in the correct locations and are correctly orientated as previously described.

Referring to Figure 8.15 wind the transformer, this is probably the most difficult part of the construction procedure and should not be rushed. Note that the diagrams do not figuratively show the required number of turns per layer. When winding the transformer take care not to over stress the bobbin otherwise pins may break

Auto electronics projects

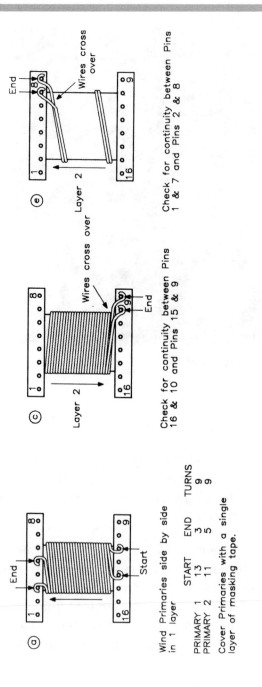

Figure 8.15 Transformer winding details

174

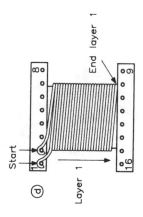

(d)

Start

Layer 1

End layer 1

| | 8 | 9 | 16 |

Wind Secondaries 3 & 4 in 2 layers
1st layer 13 turns 2nd layer 2 turns

	START	END	TURNS
SECONDARY 3	1	7	15
SECONDARY 4	2	8	15

Cover each layer with a single
layer of masking tape.

(b)

Start

Layer 1

End layer 1

| 1 | 8 | 9 | 16 |

Wind Secondaries 1 & 2 in
2 layers 13 turns each layer.

	START	END	TURNS
SECONDARY 1	16	10	26
SECONDARY 2	15	9	26

Cover each layer with a single
layer of masking tape.

Figure 8.15 Continued

Auto electronics projects

off — use pliers to carefully bend the wire around the pins. It will be necessary to remove the enamel coating from the wire to allow soldering, emery paper is ideal for this. Photo 8.4 shows an exploded view of the component parts of the transformer.

Photo 8.4   The component parts of the ferrite transformer

Cover each layer with a single layer of masking tape.

Starting at pins 13 and 11, wind bifilar two 9 turn windings of 18 SWG EC wire first, finish at pins 3 and 5 respectively, as shown in Figure 8.15(a).

Starting at pins 16 and 15, wind bifilar two 26 turn windings of 20 SWG EC wire in two layers; first wind bifilar 13 turns, as in Figure 8.15(b). Wind bifilar a further 13 turns and finish at pins 10 and 9 respectively, as in Figure 8.15(c), note the wires cross over. Check for continuity between pins 16 and 10, and 15 and 9.

Starting at pins 1 and 2, wind bifilar two 15 turn windings of 20 SWG EC wire in two layers; first wind bifilar 13 turns, as shown in Figure 8.15(d). Wind bifilar a further 2 turns and finish at pins 7 and 8 respectively, as in Figure 8.15(e), note the wires cross over. Check for continuity between pins 1 and 7, and 2 and 8.

Solder all of the leadouts to the transformer bobbin pins, fit the cores and clip into place the sprung steel core retainers.

Insert the transformer into the PCB, ensuring that pin 1 aligns with the number 1 on the PCB.

Referring to Figure 8.11 and using the 32/0.2 power connection wire, link *W number* holes; W1 to W1, W2 to W2, W3 to W3.

Finally tin the exposed lengths of PCB tracks with a *thick* layer of solder. Take care not to splash solder elsewhere which may cause short circuits.

Double-check your work and remove excess flux from the underside of the PCB using PCB cleaner. Photo 8.5 shows the assembled PCB.

Connect the two LEDs to the PCB via lengths of insulated wire as shown in Figures 8.11 and 8.16.

Testing

Figure 8.11 shows the location of the input and output connections referred to in this section.

Auto electronics projects

Photo 8.5 The assembled PCB

Fit IC1, IC2 and the fuses.

Using a multimeter on a suitable resistance range, measure the resistance between FS7 and P2, the resistance should be greater than 2 kΩ. Check also the resistance between FS2 and P2, the resistance should be greater than 2 kΩ. If significantly lower readings than stated are measured, recheck all of your work as there is likely to be a short circuit or a misplaced component.

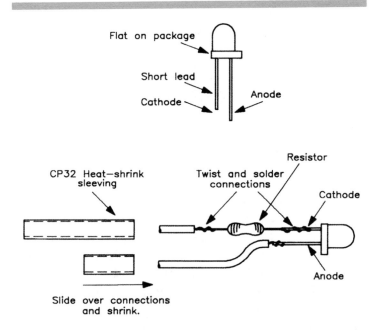

Flat on package

Short lead

Cathode

Anode

CP32 Heat—shrink sleeving

Twist and solder connections

Resistor

Cathode

Anode

Slide over connections and shrink.

Figure 8.16   LED leadout identification and connections

Connect a 12 V supply capable of delivering 5 A to the input pins P1 (+V) and P2 (0 V) via a 5 A fuse (for FS1) and a multimeter on 5 A or higher range. The quiescent current should be less than 1 mA.

Link P3 and P1 with light duty wire, whereupon the relay should energise and the power-on LED (LD1) should illuminate. The current indicated on the meter should be approximately 400 mA. If an oscilloscope and/or frequency counter are available, then these may be used to confirm that a 50 kHz (approximately) sawtooth waveform is available on P24. Avoid undue capacitive loading otherwise the frequency of the oscillator will be shifted.

Unlink P3 and P1, disconnect the supply and disconnect the multimeter.

Reconnect the supply and relink P3 and P1. Measure the voltage on the output pins, using a suitable voltage range. P4 to P7 should read +30 V with respect to P8. Pins P12 to P15 should read −30 V with respect to P8. P16 should read +12 V with respect to P17 and P18 should read −12 V with respect to P17.

The thermal shutdown circuit may be tested by carefully heating the thermistor with a hairdryer. When the thermistor reaches a temperature of approximately 80°C the thermal shutdown LED (LD2) will illuminate and the power supply will shutdown, this can be confirmed by measuring one of the supply voltage outputs. When the thermistor temperature drops to approximately 60°C the power supply will restart and the thermal shutdown LED will extinguish.

This completes testing of the power supply.

As previously stated, the power supply is specifically intended for use with two Maplin 50 W bipolar power amplifiers. In most applications the audio output power attainable from these amplifiers when used in conjunction with this power supply should be more than sufficient for in-car use. However the purist may wish to use separate power supplies for each amplifier to increase the power available per channel. Similarly, if a single channel subwoofer amplifier is required, a single amplifier may be driven from one power supply.

It is strongly recommended that the power supply is fully cased and provided with an additional external heatsink, type 2E is suggested. Metal cases are ideal for this purpose, and also provide a degree of shielding against radiated radio frequency emissions. The audio amplifiers may also be housed in the same case, which could be conveniently mounted in the car boot or under a seat. The audio amplifiers should also be heatsinked, again type 2E is suggested.

To connect the 50 W bipolar amplifiers to the power supply, treat the switched-mode power supply as a conventional power supply (as shown in the amplifier constructional details) and connect accordingly (HT1 and HT2 are positive, HT3 and HT4 are negative). Refer to Figure 8.11 for connections to the power supply. The amplifier set-up procedures should be followed in the same way as for the conventional power supply. Connections from the power supply to the amplifiers should be made using 32/0.2 wire.

Take note — Take note — Take note — Take note

Loudspeakers should be suitably rated for high power use. Beware — many car loudspeakers are given misleadingly high power ratings, so try and find out what the true r.m.s. ratings are before you use any loudspeaker. Often car loudspeaker ratings are given in peak power or total peak power, so be prepared to divide the rating by 1.414 or even 2.828! Loudspeaker wiring should also be sufficiently rated for the purpose.

Auto electronics projects

Connections from the power supply to the car electrical system should be made using very heavy duty cable. It is advisable to connect the power supply directly to the car battery via its own in-line fuse at the car battery end. Assuming a negative earth car, the chassis may be used to provide the 0 V connection, which saves on wire.

Take note — Take note — Take note — Take note

It should be pointed out that excessive sound pressure levels may lead to long term, irreversible hearing problems. High levels of sound may also blot out other external sounds, which could be dangerous when on the move. Please use common sense when using a high power in-car entertainment system.

Input:	11 to 15 V d.c., nominally 13.8 V
Input current (P_o = 116 W):	10.7 A (V_s = 11.3 V)
Output power:	120 W continuous, see note below
Outputs	
Main:	±30 V
Auxiliary:	±12 V
Continuous output current	
±30 V	2 + 2 A
±12 V	50 mA + 50 mA
Efficiency:	>90%
Thermal shut-down temperature:	80°C
Thermal shut-down hysteresis:	20°C
Standby input:	Active low
Remote switch-on input:	Active high
Thermal shut-down output:	Active low
Input noise (P_o = 120 W):	140 mV
Output noise (P_o = 120 W)	
Main:	60 mV
Auxiliary:	40 mV
Switching frequency:	25 kHz
Converter mode:	Push-pull

Table 8.1   Specification of Prototype

Car audio switched-mode PSU parts list

Resistors — All 0.6 W 1% metal film (unless specified)

R1	6k8	1	(M6K8)
R2	68 k	1	(M68K)
R3,35	1k2	2	(M1K2)
R4,5,6,7	56 Ω	4	(M56R)
R8,36	10 Ω	2	(M10R)
R9,10	1 k 3 W	2	(W1K)
R11,12	470R 3 W	2	(W470R)
R13,17,22, 23,26,27, 28,29,34	1 k	9	(M1K)
R14,19,21, 31,32,33	10 k	6	(M10K)
R15	24 k	1	(M24K)
R16,20, 24,25	4k7	4	(M4K7)
R18	1 M	1	(M1M)
R30	3k3	1	(M3K3)
TH1	15 k bead thermistor	1	(FX22Y)

Capacitors

C1,5,15, 16,25,26, 28,30	100 nF polyester	8	(BX76H)
C2,13,14	220 µF 50 V SMPS	3	(JL51F)
C3,4,11,12	1000 µF 50 V SMPS	4	(JL57M)
C6,31	2n2F 1% polystyrene	2	(BX60Q)
C7,8,9,10,17, 18,19,20	560 pF ceramic	8	(WX65V)

C21,22	1000 µF 25 V SMPS	2	(JL56L)
C23,24	10 µF 25 V tantalum	2	(WW69A)
C27	100 µF 25 V PC elect	1	(FF11M)
C29	22 µF 25 V PC elect	1	(FF06G)
C32	10 µF 16 V tantalum	1	(WW68Y)
C33,34	150 pF polystyrene	2	(BX29G)
D1,2	1N4001	2	(QL73Q)
D3,4,13, 14,15,16, 17,18	1N4148	8	(QL80B)
D5,6,7,8	BYW80-150	4	(UK63T)
D9,10, 11,12	BYW98-150	4	(UK65V)
ZD1,2	39 V BZX61C/BZX85C	2	(QF67X)
TR1	BC337	1	(QB68Y)
TR2,3,4,5	BUZ11	4	(UJ33L)
TR6,7,8, 9,10	BC559	5	(QQ18U)
LD1,2	Red LED	2	(WL27E)
RG1	µA7812UC	1	(QL32K)
RG2	µA7912UC	1	(WQ93B)
IC1	TL494	1	(RA85G)
IC2	LM311	1	(QY09K)

Miscellaneous

L1	FX4054 ferrite toroid	1	(JR84F)
L2,3	3 A RF suppressor	2	(HW06G)
T1	ETD39 ferrite core	2	(JR81C)
	ETD39 former	1	(JR82D)
	ETD39 clip	2	(JR83E)
RL1	12 V 16 A relay	1	(YX99H)

FS1	15 A 1 $^1/_4$ in AS fuse	1	(UK13P)
FS2,7	100 mA 20 mm QB fuse	2	(WR00A)
FS3,4,5,6	2 A 20 mm AS fuse	4	(WR20W)
	$1^1/_4$ in chassis fuse holder	1	(RX50E)
	fuse clip	12	(WH49D)
	6mm M3 isobolt	1 pkt	(BF51F)
	12 mm M2.5 isobolt	1 pkt	(BF55K)
	M3 isonut	1 pkt	(BF58N)
	M2.5 isonut	1 pkt	(BF59P)
	M3 isoshake	1 pkt	(BF44X)
	M2.5 isoshake	1 pkt	(BF45Y)
	M3 isotag	1 pkt	(LR64U)
	TO220 insulator	8	(QY45Y)
	T0220 bush long	1 pkt	(UL69A)
	50 W heatsink	1	(HQ69A)
	16-pin DIL skt	1	(BL19V)
	8-pin DIL skt	1	(BL17T)
	1 mm PCB pins	1 pkt	(FL24B)
	PCB	1	(GE61R)
	0.9 mm 20 SWG TC wire	1	(BL13P)
	1.6 mm 16 SWG TC wire	1	(BL11M)
	3202 green wire	1 mtr	(XR35Q)
	1.6 mm 16 SWG EC wire	1	(BL24B)
	1.25 mm 18 SWG EC wire	1	(BL25C)
	0.71 mm 22 SWG EC wire	1	(BL27E)
	lapped pair	1 mtr	(XR20W)
	CP 32 heat shrink	1 mtr	(BF88V)
	CP 16 heat shrink	1 mtr	(BF86T)
	constructors' guide	1	(XH79L)
	instruction leaflet	1	(XK50E)
	fast-setting adhesive	1	(FL45Y)

All of the above are available as a kit

switched-mode PSU kit	1	(LP39N)

Auto electronics projects

Optional (not in kit)

car fuse holder	1	(RX51F)
15 A 1 $^1/_4$ in AS fuse	1	(UK13P)
HC wire black	as req	(XR57M)
HC wire red	as req	(XR59P)
32/0.2 wire red	as req	(XR36P)
32/0.2 wire black	as req	(XR32K)
32/0.2 wire blue	as req	(XR33L)
zip wire	as req	(XR39N)
50 W power amp	2	(LW35Q)
2E heatsink	2	(HQ70M)

Everyone has a book inside them.

What's yours?

If you would like to discuss an idea for a book on a technical subject, particularly for a Maplin audience, please write a 250 word description of the book, explaining who it is for and what it is about. Send this proposal with a list of chapter headings and a description of your career and experience to date to the Publisher of the Maplin book series, Duncan Enright:

write to Duncan Enright, Butterworth-Heinemann, Linacre House, Jordan Hill, Oxford OX2 8DP, UK

or

e-mail: duncan.enright@bhein.rel.co.uk

The Publisher of Choice

MAPLBooks

These books are part of a new series developed by Butterworth-Heinemann and Maplin Electronics. These practical guides will offer electronics constructors and students a clear introduction to key topics. The books will also provide projects and design ideas; and plenty of practical information and reference data.

0 7506 2053 6 **STARTING ELECTRONICS**

0 7506 2123 0 **COMPUTER INTERFACING**

0 7506 2121 4 **AUDIO IC PROJECTS**

0 7506 2119 2 **MUSIC PROJECTS**

0 7506 2122 2 **LOGIC DESIGN**

These books are available from all good bookshops, Maplin stores, and direct from Maplin Electronics. In case of difficulty, call Reed Book Services on (01933) 414000.

Automobile Electronics
Dr E Chowanietz

A comprehensive introduction to automobile electronics, based on technology in production since the late 1980s.

As well as providing a general grounding in this rapidly developing subject, vehicle-specific applications are illustrated via a series of case studies drawn from a wide variety of different manufacturers' products. Little prior knowledge of electronic engineering is assumed, and the fundamentals of microprocessor-based systems and automotive control strategies are introduced from first principles. The text is aimed at building knowledge and developing understanding, with a minimum of jargon. It will also equip readers with the ability to understand the functioning of new automotive electronic systems as they appear on the market.

The book is ideal for degree, BTEC, City and Guilds, NVQ and GNVQ students studying motor vehicle technology, electronics or mechatronics. Motor vehicle technicians, engineers, and enthusiastic car owners seeking to keep up to date with recent developments in this fast-moving and increasingly complex field will also find it of value.

Dr Eric Chowanietz is a Principal Lecturer in the Department of Electrical and Electronic Engineering at De Montfort University, Leicester. He has industrial links with automotive companies such as Lucas and Rover Group, and has published a number of technical papers and articles on automotive sensing, monitoring and control systems.

0 7506 1878 7 320 pages Paperback

Maplin Approach to Professional Audio

An Insight into the World of Professional Audio,
From a Technical Point of View

Tim Wilkinson

This is an introduction to the sound industry from a technical
point of view. It not only provides a fascinating tour of the state
of the art, but also will encourage anyone with an interest in audio
to examine best practice and improve their own skills.

Defining the 'professional' in professional audio is no easy task, as
there is no single parameter that categorises a piece of equipment
as being particularly suited for use in a professional environment.
It is more a case of taking a look at the way in which certain types
of equipment are used in a certain environment, and to what
standards they have to conform. Based on the acclaimed series of
articles published in 'Electronics', The Maplin Magazine, this
book will give audio enthusiasts of all levels a view of the future
for audio, and ideas as to how to achieve the highest standards of
sound.

CONTENTS: Introduction; Microphones; Amplifiers; Stereo
techniques; Mixers; Analog tape recording systems; Outside
broadcast and studio recording; Digital recording; Disk-based
audio systems; Multitrack digital recorders; Index
0 7506 2120 6 288 pages Paperback

Maplin Integrated Circuits Projects

Each of the projects in this collection of data files provides a "building block" which constructors can use to experiment with components and use as a starting point for further development. This book will thus provide a toolkit for building many different types of projects and circuits based~on readily available components using straightforward techniques.

Maplin staff are experienced providers of project ideas with useful features and many applications. Each of the circuits in this book is based on components which can easily be obtained. Each project provides an excellent way of becoming familiar with the characteristics of integrated circuits, as well as providing constructional details for useful projects. The book includes relevant integrated circuit pinouts and pin designations, output waveforms, parts lists, circuit diagrams and PCB layouts of each board. The projects described here are based on those appearing in the popular Data Files features in *Electronics - the Maplin Magazine*. Some of the projects described are intended for specific applications, while many are adaptable to other applications.

CONTENTS: Over 20 integrated circuit projects of mixed types, with multiple applications.

0 7506 2578 3 192 pages Paperback